Observations on the Soviet/Canadian Transpolar Ski Trek

Medicine and Sport Science
Vol. 33

Series Editors
M. Hebbelinck, Brussels
R.J. Shephard, Toronto, Ont.

Founder and Editor from 1969 to 1984
E. Jokl, Lexington, Ky.

Basel · München · Paris · London · NewYork · New Delhi · Bangkok · Singapore · Tokyo · Sydney

Observations on the Soviet/Canadian Transpolar Ski Trek

Volume Editors
R.J. Shephard, Toronto, Ont.
A. Rode, Ottawa, Ont.

33 figures and 59 tables, 1992

Basel · München · Paris · London · New York · New Delhi · Bangkok · Singapore · Tokyo · Sydney

Medicine and Sport Science

Published on behalf of the
International Council of Sport Science and Physical Education

---

Library of Congress Cataloging-in-Publication Data
  Observations on the Soviet/Canadian transpolar skitrek / volume editors, R.J. Shephard, A. Rode.
  (Medicine and sport science; vol. 33)
  Includes bibliographical references and index.
  1. Cold – Physiological effect. 2. Circumpolar medicine.
  I. Shephard, Roy J. II. Rode, A. (Andris), 1940– . III. Series.
  ISBN–3–8055–5410–9

Drug Dosage
  The authors and the publisher have exerted every effort to ensure that drug selection and dosage set forth in this text are in accord with current recommendations and practice at the time of publication. However, in view of ongoing research, changes in government regulations, and the constant flow of information relating to drug therapy and drug reactions, the reader is urged to check the package insert for each drug for any change in indications and dosage and for added warnings and precautions. This is particularly important when the recommended agent is a new and/or infrequently employed drug.

All rights reserved.
  No part of this publication may be translated into other languages, reproduced or utilized in any form or by any means, electronic or mechanical, including photocopying, recording, microcopying, or by any information storage and retrieval system, without permission in writing from the publisher.

© Copyright 1992 by S. Karger AG, P.O. Box, CH-4009 Basel (Switzerland)
  Printed in Switzerland on acid-free paper by Thür AG Offsetdruck, Pratteln
  ISBN 3–8055–5410–9

Contents

	Acknowledgements	X

Roy J. Shephard
1 Introduction 1

Roy J. Shephard
2 Overview of the Soviet/Canadian Transpolar Ski Trek 6
   Introduction 6
   Sustained High Daily Energy Expenditure 7
   Local Exposure to Windchill 8
   General Body Cooling 9
   Ultraviolet Exposure 10
   Possible Geomagnetic Effects 10
   Psychological Stress 10
   Hormonal Changes 11
   Nutritional Consequences 11
   Conclusions 12

V.S. Koscheyev, V.K. Martens, T.S. Bashir-Zade, A.A. Belyakov, G.V. Kypor, A.V. Visochanski, M.A. Lartzev, A. Rode, M. Jetté, M. Malakhov
3 Previous Investigations of Circumpolar Stress 13
   Introduction 13
   Nature of Stress 13
      Physiological Reactions 14
      Psychological Reactions 14
   Physiological Reactions 15
   Polar Stress Syndrome 15
   Acclimatization to High Latitudes 16
      Metabolic Acclimatization 16
      Insulative Reactions 16
      Vascular Changes 17
      Respiratory Reactions 17
      Psychomotor and Psychological Changes 18
   Conclusion 18

Contents

V.S. Koscheyev, V.K. Martens, T.S. Bashir-Zade, A.A. Belyakov, G.V. Kypor, A.V. Visochanski, M.A. Lartzev, A. Rode, M. Jetté, M. Malakhov

4 Problem Formulation 19

M. Buxton, J. Sproule, M. Jetté, A. Rode

5 Personal Reflections and Medical Observations 21
Medical Observations 29

V.S. Koscheyev, V.K. Martens, T.S. Bashir-Zade, A.A. Belyakov, G.V. Kypor, A.V. Visochanski, A. Rode, M. Jetté, M. Malakhov

6 Cardiovascular Function and Autonomic Regulation 33
Methodology 33
Baseline Data 34
Observations during the Expedition 37
Observations on Completion of the Expedition 40
Relationship to Mission Success 43

M.A. Booth, J.S. Thoden, F.D. Reardon, M. Jetté, J. Quenneville, A. Rode, V.S. Koscheyev, V.K. Martens, T.S. Bashir-Zade, A.A. Belyakov, G.V. Kypor, A.V. Visochanski

7 Physical Working Capacity and Body Composition 44
Anthropometric Measurements 44
Muscle Strength and Lung Function 44
Cardiovascular Function 45
Anthropometric Data and Muscle Strength 46
Cardiovascular Data Obtained on Canadian Team 48
Ergometric Data from Moscow 55
    Initial Results 55
    Final Results 56
Correlation with Expert Appraisals 58

B.A. Utehin, M.G. Malakhov, M. Jetté, A. Rode, S. Livingstone, R.W. Nolan, A.A. Keefe

8 Chamber Simulation of Ski Trek 59
Introduction 59
Methodology 59
    Soviet Studies 59
    Canadian Studies 60
Results 61
    Soviet Studies 61
    Canadian Studies 62
    Changes in Thermal Reactions 63
    Reactions of Cardiorespiratory and Central Nervous Systems to
        Soviet Chamber Simulations 67
    Relation to Mission Success 70

Contents

V.S. Koscheyev, M.A. Lartzev, A. Rode, M. Malakhov

9 Psychic Adaption of Participants ... 71
General Considerations ... 71
    Overall Effects of Stress ... 71
    Stresses Associated with High Latitudes ... 71
    Psychic Adaptions to Stress ... 72
    Implications for Ski Trek ... 73
Methods for the Study of Psychic Adjustments ... 74
    Clinical Approach ... 74
    Use of Standard Questionnaires ... 74
    Subjective Assessments ... 75
    Other Test Procedures ... 75
Experimental Plan ... 76
Initial Status of Participants ... 76
Comparison of Soviet Team with Participants in Previous Expeditions ... 79
Characteristics of Canadian Participants ... 81
Comparison of Soviet and Canadian Participants ... 83
Data Obtained During and Following the Trek ... 84
    First Drop ... 84
    Findings at Third Drop ... 86
    Findings at Fourth and Fifth Drops ... 87
    Findings after Completion of the Trek ... 87
Subjective Appraisals ... 88
    Methodology ... 88
    Initial Data ... 88
    Assessments during the Trek ... 88
    Final Data ... 89
Relationship of Appraisals to Mission Performance ... 89
    Profile of Successful Participants ... 91
    Profile of Least Successful Participants ... 91
    Observations during the Traverse ... 93
    Observations on Completion of Trek ... 93
    Comparison of Best and Worst Participants ... 95
Underlying Motivations ... 96
Overall Analysis of Behavior ... 97
Group Dynamics ... 99
Identification of Successful Participants ... 100
Conclusions ... 102
Appendix A ... 103
Appendix B ... 104

R.A. Tigranian, N.F. Kalita, N.A. Davydova, B.R. Dorokhova, M.G. Malakhov, A.G. Melkonian, A.S. Roganov, I.D. Stalnaya, T.I. Chernikhovskaya, T.N. Shumilina

10 Status of Selected Hormones and Biologically Active Compounds ... 106
Introduction ... 106
Methodology ... 107
Sympathoadrenal, Serotoninergic and Histaminergic Systems ... 108

General Considerations	108
Changes during the Expedition	109
Conclusions	114
Pituitary/Adrenocortical Axis	115
General Considerations	115
Changes over the Course of the Expedition	115
Renin/Angiotensin/Aldosterone Axis	118
General Considerations	118
Changes Observed during the Expedition	119
Hypothalamus/Hypophysis/Gonadal Axis	123
General Considerations	123
Changes over the Expedition	123
Hypothalamus/Hypophysis/Thyroid Axis	124
General Considerations	124
Changes Observed during the Expedition	125
Hormone Ratios	125
Conclusions	127
Somatotropin, Somatostatin and Somatomedin-C	127
General Considerations	127
Changes Observed during the Expedition	128
Interactions with Other Hormones	129
Parathormone/Calcitonin System	129
General Considerations	129
Changes over the Course of the Expedition	129
Insulin/C-Peptide/Glucagon System	130
General Considerations	130
Changes during the Expedition	130
Conclusions	131
Neuropeptide Concentrations	131
General Considerations	131
Changes during the Expedition	132
Biologically Active Compounds	134
General Considerations	134
Changes during the Expedition	134
Stress, Carbohydrate and Lipid Metabolism	135
General Considerations	135
Changes during the Expedition	136
Evidence of Stress and Adaptive Responses	137

L.E. Panin, N.M. Mayaskaya, A.A. Borodin, P.E. Vloshinsky, I.E. Kolosova, A.R. Kolpakov, V.G. Kunitsin, M.F. Nekrasova, N.G. Kolosova, L.S. Ostanina, T.A. Tretyakova

11 Comparison of Biochemical Reactions to Trek and Chamber Simulations ... 139

Introduction	139
Glucose-Regulating Hormones	139
Methodology	139
Responses before the Expedition	139

1 Introduction
Roy J. Shephard

The completion of an 1,800-km cross-country ski trek in a period of 3 months is a notable feat under any environmental conditions, deserving the detailed consideration of the physician, the exercise scientist and the dietician. At a minimum, the body is exposed to 8 h of continued and demanding physical activity for each of 91 consecutive days. Interest must be particularly high when some of the participants are in their mid or late forties, and the route covers the rough territory of the polar ice cap, with no support except for occasional drops of limited rations from an aircraft. In such circumstances, the participants face not only heavy physical stresses, but also exposure to severe cold, severe ultraviolet irradiation, potential deficiencies of key vitamins, and the psychological stresses associated with unknown dangers. This in essence is the scenario of the joint USSR/Canadian transpolar ski trek of 1988, which forms the theme of this monograph.

North American and Western European literature is only now beginning to explore changes of body function induced by various forms of stress, including exercise and very cold environments [Shephard, 1991]. To date, it appears that a moderate amount of stress has a positive effect upon most individuals, but that heavier stress has a depressant effect, upsetting the normal balance of anabolic and catabolic hormones. There has been little study of the impact of combined stresses such as exercise, cold, excessive ultraviolet irradiation, and exposure to danger, but from our current understanding it may be hypothesized that adverse reactions are likely to occur with a lower dose of exercise when this is presented in combination with other stresses. Such information has a multitude of practical applications. The base data on negative reactions to excessive stress could explain why so many athletes succumb to acute infections immediately before a major competition. A careful monitoring of various aspects of psycho-physiological and endocrine function might provide an early warning that an athlete was overtraining. Optimization of the stress response might also find therapeutic application in the treatment of autoimmune disorders, cancer and immune deficiency syndromes, possibly pointing out a need for dietary supplementation and/or use of anabolic stimulants. The activities involved in the trek-prolonged skiing and cold survival have their lesser parallels during mountaineering expeditions and recreational skiing. Finally, the results could have more everyday applications in improving the lives of workers who are attempting to exploit the circumpolar regions. Such individuals also face long periods when a limited, vitamin-poor diet prepares them poorly for an environment where the peripheral circulation is challenged by physical activity in extreme cold. Sunburn can increase the risk of cutaneous neoplasms, while the emotional stresses of an unfamiliar and occasionally dangerous environment can upset the normal delicate hormonal balance governing anabolism and catabolism.

The participants in the trek were nine Soviets and four Canadians. The two largest countries in the world, the USSR and Canada, both have a substantial stake in the development of their northern territories and in the bringing of 'southern' amenities to the indigenous peoples of these regions. Soviet investigators have long been intrigued by interactions between physiological stresses, psychological reactions and humoral responses. Unique methods of evaluating metabolic stress have been evolved by Soviet scientists, including both a study of hormonal ratios, and also a detailed analysis of membrane properties. It is thus particularly fortunate that the expedition was heavily weighted by Soviet scientists. Canadian investigators also examined some of the traditional physiological measurements, providing detailed data on the energy costs of cross-country skiing and its relation to overall cardiovascular health, and keeping detailed records of medical problems, particularly frostbite and sunburn.

For many Western readers, much interest will attach to the thinking of the Soviet scientists, since their work is not normally published in the English language. The present monograph thus concentrates on the expedition as perceived by the Soviet team. Where possible, a direct translation of their text has been used, although when appropriate the editors have interpolated notes and paraphrases to make the ideas more accessible to occidental scholars.

The Soviets see the development of the circumpolar regions as a current priority of several governments with responsibility for the arctic territories. In support of this view, one may note the growing attendance at the triennial conferences on circumpolar health. At the most recent of these gatherings, held in Whitehorse (Yukon Territory) in the summer of 1990, more than 1,000 investigators met to consider many aspects of the arctic habitat. A first need in providing medical care and optimal working conditions for the development of cities at high latitudes is to solve some of the varied problems of human acclimatization to life in such regions. We may note in passing that the Soviet scholars commonly refer to *adaptation* to cold, in keeping with their concepts of stress adaptation; however, in Western thought a distinction is drawn between: habituation (a process of becoming accustomed to the cold and having a lesser subjective reaction to it); acclimation (the process of adjustment that develops through a succession of controlled exposures in a climatic chamber); acclimatization (the greater process of adjustment that occurs as a person is exposed to the totality of the arctic habitat) and adaptation (a process of genetic change which fits particular groups, possibly including the Inuit, to colonization of a harsh habitat. The particular stresses noted by the Soviets as affecting human residents of high latitudes (whether arctic or antarctic) include a very cold climate, exposure to unusual cycles of light and darkness, dietary restrictions and exposure to an unusual geomagnetic field.

The ski-trek party were certainly exposed to quite low environmental temperatures (a mean of $-38\,°C$ in March, $-33\,°C$ in April, and $-11\,°C$ in May) for a prolonged period, although they were well clothed, and during the daytime they were working quite hard, so that the impact upon core temperature may not have been very great. In the Soviet Union, industrial workers in the north are apparently exposed to low temperatures for quite long periods, and during the 1978 Circumpolar Health Conference in Novosibirsk, a long session that the senior editor chaired was concerned with the permissible hours of outdoor labor for men constructing a new northern Siberian railway (the BAM Railway). There are some groups in Canada, also, who are exposed to severe cold, but many people who work in the north spend much of their time in heated shelters or are so well clothed that the main problem they encounter is hyperthermia rather than hypothermia during outdoor activity.

The immigrant to the north certainly finds both the long nights of the winter and the lack of darkness in the summer stresses to which an early adjustment must be made. The ski-trek party were travelling at a period when the hours of daylight were relatively normal, although because they were travelling outdoors in clear air for long periods, with much reflected radiation from snow and ice, their exposure to ultraviolet radiation was greater than normal.

Since the ski-trek party carried their own food, their diet was severely restricted. In some northern communities, the variety and vitamin content of food is still much less than in the south, although in Canada the main restriction upon the diet at high latitudes is now budgetary, rather than logistic, and in most communities the standard 'southern' diet is available to anyone willing and able to pay airfreight. A lack of fresh fruit and vegetables leaves some northern residents, particularly the indigenous peoples, with less than desirable blood levels of ascorbic acid [Nutrition Canada, 1967], while a combination of ultraviolet light and hard physical work may conspire to produce deficiencies of α-tocopherol (vitamin E).

The extent of any geomagnetic effects upon the human body remains highly controversial. While some Soviet investigators have described significant influences from such exposure [Kaznacheev, 1974], sceptics have argued that the geomagnetic changes were not always clearly separated from other stresses inherent in the arctic milieu. In North America, various investigations have been conducted on those working on or living near high-tension power lines. The majority of these studies have found no effect from the altered electromagnetic field, but critics of such reports have been quick to point out that the sponsor has usually been the company supplying electricity to the region.

The main objectives of the applied psychologist, the applied physiologist and the ergonomist are normally to protect the individual worker against the stresses of all types that are encountered in extreme environments, thereby normalizing the productivity and increasing the health and job satisfaction of northern workers. Protection of the individual employee from excessive stress is based upon a closely integrated combination of social, scientific, engineering, ergonomic and medical measures, all designed to maximize the occupational adjustment of humans who work in ecologically unfavorable environments. Where possible, the goal is to avoid mandatory requirements for the wearing of personal protective clothing, since this increases energy expenditures and reduces dexterity.

The support of the dietician, the applied psychologist, the applied physiologist, the pharmacologist and the ergonomist is particularly important to members of scientific and geological expeditions such as the ski-trek participants, to military units, and to other individuals who must undertake long periods of vigorous physical work outdoors. The Canadian observations point out the need for protection of the extremities and augmentation of the peripheral circulation when older people are active in the arctic; they also highlight the need for effective blocking agents against excessive ultraviolet irradiation. The Soviet authors have suggested that specific issues requiring further study include: (i) the dependence of adaptive reactions on the individual's initial functional status; (ii) the social, interpersonal and psychic adaptations that occur among participants in multinational expeditions; (iii) interrelationships between physiological and psychological adaptations, and (iv) details of the biochemical and immunological adaptations during arduous outdoor activity under extreme northern conditions.

The transpolar ski-trek was initiated by the Siberian Branch, Soviet Academy of Medical Sciences, the USSR Ministry of Public Health and by Canadian Scientists with the objective of studying these various issues. The specific objectives included: (i) an evaluation of the functional status of participants before, during and after completion of the trek; (ii) a study of psychological adaptations occurring over the course of the trek; (iii) a detailed review of biochemical and humoral adaptations, and (iv) a study of changes in the intestinal flora and immune function.

Test measurements were selected to explore a broad range of variables relevant to these objectives. Baseline data were collected in Moscow, in Ottawa, and at the experimental station of the Novosibirsk Institute of Biochemistry. Further studies were made at 2 week intervals throughout the transpolar crossing, particularly at a point near the North Pole, when the participants had covered two thirds of their journey. Final observations were made on reaching Canada, both in the high arctic and in Ottawa. Much of the baseline and final data was collected in climatic chambers under conditions that simulated as far as possible the intensity, duration and type of stresses anticipated during the actual trek. A partial listing of measurements includes the following observations.

Cold exposure: Observations on climate and development of frostbite; determinations of environmental temperature and wind speeds; regional thermal discomfort assessments; the measurement of skin temperatures at five points, as described by N. Witte; the measurement of rectal temperatures, and the calculation of a weighted average skin temperature and an average body temperature; cold immersion tests; assessments of clothing insulation.

Clinical data: personal observations on mood state and experiences, together with detailed records of medical problems encountered during the trek.

Cardiorespiratory function: resting systemic arterial pressure, 12-lead electrocardiography, and analyses of variations in pulse pressure and cardiac rhythm; ballistocardiographic estimates of stroke volume, myocardial contractility and cardiac output; determinations of respiratory rate and respiratory minute volume; determinations of the physical working capacity at a heart rate of 170 beats/min and maximal oxygen intake; measurements of the mechanical efficiency of treadmill skiing, with and without a backpack; heart rate estimates of daily energy expenditure during the trek.

Nutrition and anthropometry: height, body mass, skinfold predictions of body fat and lean body mass, measurements of shoulder and grip strength.

Psycho-physiological function: a simple visuo-motor reaction time, an integrated visuo-motor reaction time, and reactions to a light stimulus.

Psychosociological Function: the Minnesota Multiphasic Personality Inventory; Cattell's 16-factor personality test; Raven's Progressive Matrices; Lucher's color test; Differentiated self-appraisal method; expert appraisal; subjective appraisal of personal qualities; a sociometric questionnaire.

Biochemical and immune function: detailed examination of blood hormone levels, carbohydrate, protein and fat metabolism, lipid profile, erythrocyte glycolysis, and properties of the erythrocyte membrane.

2 Overview of the Soviet/Canadian Transpolar Ski Trek
Roy J. Shephard

Introduction

A combined Soviet/Canadian team of 13 young and middle-aged men completed a transpolar ski trek over a total estimated distance of 1,730 km, from Cape Arkticheski on the northern tip of Siberia's Severnaya Zemlya Islands to Cape Columbia in the northern-most part of the Canadian arctic. The objectives of the expedition were partly political (developing international collaboration and establishing sovereignty over the polar ice cap), and partly scientific, to evaluate the responses of adult males to the combined stresses of prolonged cold exposure and intensive physical activity.

The overall leadership of the expedition was provided by D. Shparo, with several participants leading different sections of the route. The Soviet participants included a pulmonary physician, Dr. Mikhail Malakov, and all of their group had extensive experience of similar circumpolar exploits. The Canadian team included a general practitioner and competitive cyclist, Dr. Max Buxton; a long-distance runner, Rev. Laurie Dexter; a nation flat-water canoe champion, C. Holloway, and a national cross-country ski champion, R. Weber. While the Canadian team had a high level of aerobic fitness, the Soviet participants had the advantage of much greater experience in skiing and in the endurance of arctic conditions.

As a part of the scientific objectives, simulations of the trek were undertaken in climatic chambers in Moscow (before the trek), and in Ottawa (immediately after completion of the trek).

The traverse was accomplished with no outside logistic assistance other than occasional drops of food, fuel, batteries and replacements for broken equipment from Soviet and Canadian aircraft. The timing was from March 3 to June 1, 1988, under ground and weather conditions that were at times quite arduous. The total journey occupied 91 days, so that the participants were required to ski for an average of just under 20 km/day, carrying not only their immediate supplies but also substantial amounts of scientific equipment in heavy back-packs.

This brief overview of the mission considers the specific stresses faced by the participants including: (i) a sustained high daily energy expenditure; (ii) local exposure of the face, extremities and airways to severe windchill;

(iii) occasional general body cooling; (iv) extensive exposure to ultraviolet irradiation; (v) possible geomagnetic effects; (vi) psychological pressures; (vii) hormonal consequences of sustained exposure to both physical danger and psychological pressures, and (viii) nutritional consequences of a restricted, high-fat diet.

Sustained High Daily Energy Expenditure

In theory, cross-country skiing is a relatively efficient method of transportation for a human who wishes to cover a long distance over snow and ice, particularly when a heavy load such as the 37- 45-kg pack of the transpolar trek must be carried [Shephard, 1982]. However, the actual energy cost depends greatly on the skiing speed, terrain, snow and ice conditions, and the experience of the skier. Booth et al. [1989] estimated the skiing speed and work rate of the participants by comparing laboratory treadmill skiing speeds and heart rates with field measurements of heart rate. Their data equated to a 3.5 km/h speed [chapter 6].

During the actual trek, the skiers rested 10 min/h. After 4–5 such legs, a 1-hour rest was taken to prepare and eat a meal. The team were active for a total of 8–10 h/day at a relative stress that was estimated at 39–46% of maximal oxygen intake. Nights were spent in sleeping bags, with additional protection from an unheated nylon tent (designed to hold 9, but actually accomodating 13 people). Some members of the team eventually became frustrated by condensation and lack of space inside the small tent, and resorted to building snow shelters for their protection at night.

The backpack that was carried during the day increased each individual's overall mass by an average of 50%, but laboratory tests suggested that energy expenditures rose by an average of only 36%. In addition to the primary task of skiing, substantial amounts of physical work were accumulated during such duties as setting up and striking camp, cutting ice and preparing meals each day. Although the total daily energy expenditure was moderate relative to a single sustained athletic competition such as a triathlon, at a rate of 5–7 kcal/min (21–29 kJ/min) for much of an 8-hour day, it would be regarded as very arduous industrial work. The total was likely 17–21 MJ (4,000–5,000 kcal/day).

Jetté et al. [1989] observed a 15–20% decrease in predicted maximal oxygen intake among the Canadian participants over the course of the traverse [chapter 6]. They argued that the Canadian team had trained very hard in preparation for the crossing, and that increases in skiing efficiency averaging 9% contributed to an intensity of daily work during the trek that was insufficient to sustain their physical condition. The initial readings for aerobic power, averaging 62 ml/kg · min for the 4 Canadian participants, were very high, reflecting their initial experience of cross-country skiing and other long-distance sports. Others, such as Durnin et al. [1960] have found that prolonged moderate exercise is quite an effective method of training. It might thus be argued that the apparent loss of physical condition reflected difficulties in matching ergometer calibration between initial and final measurements. In support of this view, the initial average peak power output of the Soviet participants, as measured on the Moscow ergometer [chapter 6], some 265 W, implies a $\dot{V}_{O_2 \, max}$ of only about 45 ml/kg · min, comparable with the 42 ml/kg · min observed after the trek [Jetté et al., 1989]. On the other hand, the Canadians who showed the decrease in aerobic power were, in general, the fittest members of the group in terms of aerobic power, and they were tested in the same laboratory by the same observers, using the same protocol before and after the expedition. Moreover, the average predicted values for 3 of the Canadians (60.8 versus 61.4 ml/kg · min) agreed quite closely, both between Moscow and Ottawa, and with the direct measurements. However, one lesson for future studies would be to exclude any possibility of an altered calibration factor by making a biological calibration of the equipment before and after the journey, using observers not directly involved in the trek [Jones and Kane, 1979].

The reported loss of physical condition, if distributed across all subjects, might also seem at variance with the gains of strength and lean mass [chapter 6] (although an accumulation of muscle can sometimes reduce relative aerobic power). The loss of physical condition thus remains a somewhat controversial finding. Although subjects did not develop severe cumulative fatigue, muscular aches and pains were relatively common during the trek [Sproule et al., 1989]. However, there is no strong evidence that the required effort was excessive for this well-trained group. Cathepsin levels decreased over the trek [Panin et al., 1990b; chapter 10], and indeed some of those involved in the event thought that the daily dose of exercise was too light rather than too heavy.

Local Exposure to Windchill

The polar temperature in early March can be as low as -50 °C, and for the first 6 weeks of the trek the temperature rarely exceeded -40 °C. With full arctic clothing, only the face and airways are exposed to windchill. Frostbite of the extremities can reflect failure to wear appropriate protective equipment (for instance, the removal of gloves to facilitate completion of a specific task), a loss of insulation due to an accumulation of sweat within the clothing or water immersion, or poor clothing design (for instance inadequate closure of ankles and wrists, a poorly designed hood, or insufficient insulation of the feet). All 13 members of the trek suffered minor forms of cold injury [Sproule et al., 1989; chapter 11], but in all except 2 individuals the problem had resolved by the time of examination in Ottawa [chapter 11]. As might be expected, the severe cold exposure of the extremities induced a local cold acclimatization [Livingstone et al., 1990; chapter 7].

The two severe cases of frostbite (one on the nose, the other on the great toe) led to some tissue loss, and sensory nerves to the affected toe were also damaged [Sproule et al., 1989; chapter 11].

The provocation of an acute bronchospasm by inspiration of cold, dry air is well recognized [Morton and Fitch, 1990]. There has also been vigorous discussion on the development of more permanent respiratory lesions as a result of prolonged exposure of the airways to very severe cold [Hildes et al., 1976; Rode and Shephard, 1985]. Some reports have described impaired static and dynamic lung volumes, dilatation of the pulmonary vessels, and ECG right bundle block, while others have found very normal cardiorespiratory function in circumpolar populations. Three of the 13 participants in the trek showed some evidence of mild restrictive lung disease after completing the expedition [Sproule et al., 1989; chapter 11], but the poor lung function scores could possibly reflect fatigue or an acute hypersecretion of mucus rather than any more permanent pathology.

General Body Cooling

The clothing worn on the trek was tested by cold chamber exposure of volunteers at $-20\ °C$ [chapter 7]. These observations showed that both Canadian and Russian clothing kept the core temperature within acceptable physiological limits, although the Canadian outfits apparently provided more effective thermal insulation, particularly when the subjects were working hard [Utehin and Malakhov, 1990].

Comparison of cold responses before and after the trek suggested the development of an insulative reaction, with a reduction of resting metabolic rate and delayed shivering [Livingstone et al., 1990; chapter 7]. This suggests that participants experienced some general cold stress, although in the chamber simulations there was only a minimal drop in average body temperature. The reports of an insulative reaction are at puzzling variance with the local vasodilatation, as evidenced by less digital cooling in iced water.

Ultraviolet Exposure

The effects of clear arctic air and long periods of outdoor activity lead to an intense exposure to ultraviolet light, with possible peroxidation of fats [chapter 10]. Additional ultraviolet exposure is produced by reflection from snow and ice, giving a relentless glare. Sproule et al. [1989] noted 8 individuals who had skin affectations and 7 with eye irritation from this source [chapter 11]. Such irritation is a concern, since prolonged exposure can give rise to keratopathy, pterygium and pinguecula [Forsius, 1976].

In the early stages of the trek, there were signs of lipid peroxidation [Kaznacheev, 1976; chapter 10], but this later disappeared, perhaps due to a compensatory increase in superoxide dismutases.

Possible Geomagnetic Effects

There has been much speculation as to possible geomagnetic effects on the body from the strong polar attraction [Kaznacheev, 1976]. Epidemiological studies of hydro workers and of those living near major transmission lines have tended to discount such responses. Kaznacheyev [1976] has suggested that various physiological variables are associated with the strength of the geomagnetic field (urinary excretion of 17-ketosteroids and adrenaline, choline esterase activity, urinary vitamin B_1, weighted skin temperature, blood flow rate, cardiac output, blood pressure, pulse pressure, hemoglobin concentration, blood oxygen capacity and erythrocyte sedimentation rate. However, a multiple regression analysis would be needed in order to be sure that other features of polar exploration such as cold exposure, disturbed circadian rhythms, isolation and psychological stress were not responsible for these effects.

Psychological Stress

The psychological stresses imposed on the participants [chapter 8] included difficulties of communication between the two teams, differences of religion, language and culture, a combination of isolation and severe overcrowding (13 men packed into a tent 4 m in diameter), and an adverse environment with substantial physical danger (potential attacks from polar bears, one participant almost drowned in a fault in the ice, various physical injuries and part of the tent destroyed by fire from a malfunctioning stove).

The problems of interpersonal relationships between the two teams seemed dominant [Koscheyev et al., 1990]. No very clear profile of the most effective participants emerged from the preliminary physiological and psychological data [Bshir-Zader et al., 1990; chapters 6–8], although there was some correlation between consistency of resting heart rates (a measure of sinus arrhythmia suppression by the sympathetic nervous system), consistent reaction times and overall mission success.

Hormonal Changes

Hormonal reactions might be anticipated to: (i) prolonged exercise (increased output of β-endorphins, growth hormone, catecholamines, decreased output of sex hormones, diabetic-type and over-training reactions); (ii) sustained cold exposure (modifications of fluid regulating hormones and metabolism regulating hormones); (iii) psychological stress (ACTH, cortisol levels), and (iv) high levels of ozone.

The data collected by the ski-trek team were generally in keeping with these expectations [chapters 9, 10]. Exercise responses included a biphasic change in concentrations of β-endorphins and substance P [Tigranian et al., 1990], increased blood levels of growth hormone and somatostatin but decreased somatomedin [Dorokova et al., 1990], increased blood levels of epinephrine and norepinephrine and serotonin [Davydova et al., 1990], a loss of glucagon response to exercise over the trek (a diabetic-type reaction) [Panin et al., 1990d], reduced pituitary-gonadal function [Melkonian et al., 1990], increased blood and urinary cAMP, and increased PgE and prostacyclin [Shumulina et al., 1990]. Cold-related responses included substantial changes in blood levels of angiotensin II, acetylcholine esterase and antidiuretic hormone [Roganov et al., 1990], and somewhat surprisingly decreased thyroid function [Kalita et al., 1990a], while stress reactions (ACTH and cortisol) were seen before departure in Moscow, with continued high cortisol levels over the first half of the trek [Kalita et al., 1990b]. Finally, as noted above, there were suggestions of adaptive increases in superoxide dismutases [Panin et al., 1990a].

Nutritional Consequences

Diet was limited by problems of supply, cooking, and the need for a high daily energy intake. General clinical observations showed short episodes of abdominal cramps and diarrhea [Sproule et al., 1989; chapter 11], and this was attributed to the high-fat content of the diet [chapter 10].

Depot fat mobilization was increased by the high levels of cAMP, and there was evidence that the proportional usage of fat increased over the courses of the trek. In the Soviet team members (who had a somewhat higher percentage of body fat than the Canadian participants), body fat decreased from 24.5 to 19.4% on the skinfold data and from 20.9% to 16.1% by underwater weighing over the course of the trek [chapter 6]. These factors contributed to development of what the Soviet investigators regard as a pre-diabetic state [Panin et al., 1990c; chapter 9]. Examination of the chain of glycolysis [Panin et al., 1990a, c] suggested that the rate-limiting step in glycolysis, normally located after the enzyme phosphofructokinase, was displaced further down the metabolic chain by the combined exercise and cold stress [chapter 10]. There was a corresponding increase in levels of high-density lipoprotein-2 cholesterol over the course of the trip, providing further evidence that participants performed a large volume of work during the traverse [chapter 10]. Other data showed substantial alterations in the properties of the erythrocyte membrane, possibly a local adaptation to increase oxygen transport, or possibly a more general disturbance of membrane function [chapter 10].

Conclusions

Well-trained young and middle-aged adults from an urban environment were able to participate in a well-planned but prolonged arctic winter trek without serious adverse consequences for either physiological or psychological systems. Nevertheless, the stresses encountered during the mission did seem to stimulate acclimatization through an activation of the autonomic system. During such feats, the body is severely stressed, and it is important to regulate the intensity of such stressors if over-stress and deregulation of the normal systems of acclimatization and adjustment are to be avoided.

3 Previous Investigations of Circumpolar Stress

V.S. Koscheyev, V.K. Martens, T.S. Bashir-Zade,
A.A. Belyakov, G.V. Kypor, A.V. Visochanski, M.A. Lartzev,
A. Rode, M. Jetté, M. Malakhov

Introduction

Western studies of adaptation to the circumpolar habitat can be found in the Conference on Polar Human Biology sponsored by the Scott Polar Institute [Edholm and Gunderson, 1973] and the published proceedings of several Circumpolar Health Conferences [Shephard and Itoh, 1976; Hart-Hansen and Harvald, 1981; Fortuine, 1985; Linderholm, 1989]. An extensive bibliography of Soviet Research in this domain is also available [Kaznacheev et al., 1985].

The overall human impact of exposure to high latitudes is likely to be felt most acutely when expeditions traverse the unpopulated areas of the Arctic and Antarctic with little logistic support. However, until quite recently, Soviet medical studies, like most of their Western counterparts, have been limited to tractor/trailer expeditions undertaken from the shores of Antarctica to the interior research stations [Volovich, 1983; Deryapa, 1965]. Such studies have suggested that there are significant differences in polar reactions between those wintering in the coastal regions of Antarctica and those experiencing the added stress of hypoxia in continental areas of the Antarctic [Arnoldi, 1962; Burton and Edholm, 1957; Produtskij and Vorobyev, 1969].

Nature of Stress

Various previous investigators from the Soviet Union have defined the typical circumpolar climate as uncomfortable and severe [Arnoldi, 1962; Danishevskij, 1968; Kandror, 1968; Kaznacheev, 1974] or even as extreme [Baevskyj, 1974]. Irrespective of the description, it is agreed that the nature of the environment places a severe stress upon physiological, psychological and sociopsychological function, leading to a deterioration of working efficiency, both acutely and during periods of prolonged residence [Semenov, 1976; Shchedrin, 1976; Kaznacheev et al., 1976, Tarkhan, 1976].

Stresses may arise not only from the direct effects of the harsh natural environment itself, but also from the unusual sociopsychological environment and occupational demands faced by those living and working in the north. The mechanisms of acclimatization developed by the body are, moreover, closely interlinked, so that a disturbance of acclimatization in one domain (for example, the psychological) could have important consequences for function in some other system (for example, the physiological response to exercise).

Physiological Reactions

Soviet investigators specializing in cold exposure and occupational physiology have already collected copious data on the changes that are observed when subjects are exposed to the extreme environments of high latitudes for short and for longer periods. Previous studies have documented changes in cardiovascular, respiratory and other physiological systems, and have made a detailed analysis of related biochemical and immunological reactions [Burton and Edholm, 1957; Bobrov et al., 1979; Ilyin, 1971; Kaluzhenko, 1971; Poggenpol and Ilyin, 1971].

Psychological Reactions

Perhaps because of the strong psychophysiological emphasis of many Soviet laboratories, the attention of Soviet researchers has inevitably been drawn to the psychic component of reactions to high latitudes, particularly the overall response to the stresses imposed by a combination of a severe climate, life in a small and remote settlement, the monotony of the natural environment (often with long spells of grey-out and white-out), the resultant partial social and sensory deprivation and limited opportunities for communication with the outside world. All of these adverse factors place severe demands on the neuropsychic sphere of human activity.

The study of human reactions in such a milieu has potential relevance not only to the immediate problem of transpolar expeditions, but also to the staffing of research stations in the arctic and antarctic, together with the more general occupational problems encountered during prolonged sea and space voyages. However, because of differences of motivation between different categories of northern residents, the patterns of adjustment seen in one group cannot be transferred uncritically to analysis of findings in an entirely different population.

Physiological Reactions

The physiological reactions to polar stresses have previously been studied mainly on small parties of scientists who were involved in tractor/sledge expeditions, on staff residents at polar research stations, and on individuals living and working in polar settlements [Volovich, 1983; Deryapa, 1965; Shechenko et al., 1976]. The Canadian Defence Laboratories have made some measurements of energy expenditures and fat losses during 5- to 10-day military sledging expeditions from Fort Churchill [O'Hara et al., 1978], and the energy expenditure of Inuit hunters has also received limited investigation [Godin and Shephard, 1973]. The British Antarctic Expedition presented a substantial verbal account of their work at the 1984 Circumpolar Health Conference in Alaska, although this information was unfortunately not included in the published proceedings. Soviet scientists have to date published almost no papers in the Western literature on the physiological problems encountered by field staff, military expeditions and hunters, although they are agreed that such groups experience the greatest hardships during their life and work in the far north.

Polar Stress Syndrome

One important feature of early human acclimatization to high latitudes is an adjustment to the differing length of the polar day and night [Danishevskij, 1968]. During this early phase, the individual often develops what has been termed the polar stress syndrome, a specific human response to a complex array of social, psychological and biophysical disturbances. The biophysical disturbances specifically affect cell membranes. Soviet scientists have argued a causal role for the geomagnetic and cosmic influences that are particularly strong in the circumpolar regions.

The polar stress syndrome is associated with both neurophysiological and biochemical disturbances, affecting homeostasis in neuropsychic, somatic and autonomic systems. There is a marked desynchronization of normal diurnal rhythms [Baevskyj, 1974; Bazhenov et al., 1974], with a decline in the stability and a narrowed range of possible variations in neuropsychic, motor and autonomic functions [Bobrov et al., 1979]. Parasympathetic tonus is dominant [Kulikov and Lyakhovich, 1980], but nevertheless there is a decrease in physical working capacity, a loss of accuracy in the performance of complex sensorimotor tasks [Kaluzenko, 1971], an activation of free radical oxidation, changes in the levels of enzymatic and non-enzymatic antioxidants [Kaznacheev, 1976] and an activation of nitrogenous lipid metabolism [Deryapa and Ryabinina, 1977].

Some of these potential disturbances are explored in chapter 10 of this report.

Acclimatization to High Latitudes

Physiological aspects of long-term adjustment to life in the arctic and antarctic include changes in thermoregulation, physical efficiency, hematological measures and biochemical responses [Danishevskij, 1968; Deryapa and Ryabinina, 1977; Curtis, 1985]. Regulatory changes may be differentiated into metabolic and insulative reactions [Shephard, 1985].

Metabolic Acclimatization

The metabolic type of acclimatization is associated with an increase in energy consumption when performing a given activity in a given environment. The augmented metabolic rate allows a greater flow of heat to the skin surface and a negligible change in skin temperature despite the cold environment. The metabolic pattern of reaction is also characterized by a stable body core temperature.

Heat production is increased [Ivanov, 1972; Khaskin, 1975] through a combination of nonshivering thermogenesis and an increase in muscle metabolism (either as an increase in tonus or as frank shivering). Changes in the external environmental temperature result in an immediate flow of impulses from thermoreceptors to the central nervous system, triggering sympathetically driven metabolic responses that allow a greater peripheral blood flow, maintenance of skin temperature, and conservation of manual dexterity.

Insulative Reactions

Insulative reactions, in contrast, restrict heat loss from the skin surface, with a decrease in peripheral blood flow, a drop in skin temperature and the suppression of sensations of coldness.

The insulative type of reaction is economical, in that it minimizes the need for heat generation, but the body core temperature is less stable than with a metabolic type of reaction. Insulative reactions are particularly characteristic of humans who are exposed to a combination of cold air and high wind velocities [Panin et al., 1979]. Studies of Europeans who have spent a winter at Antarctic stations suggest that the insulative reaction is also more frequently encountered in the high continental regions of Antarctica [Poggenpol and Ilyin, 1971; Tikhorimov, 1968; Boriskin, 1973] than in the coastal zones, possibly because peripheral vasoconstriction is a feature of high altitude hypoxia. There is an associated slowing of the frequency of respiration, without an increase in tidal volume [Skorobogatova et al., 1969]. This last feature of acclimatization reduces the water loss in the expired air (and thus respiratory cooling) by a factor of 1.4 [Yakimenko and Simonova, 1977].

Vascular Changes

Cold acclimatization is usually marked by a less severe decrease in skin temperature when the individual is exposed to any given intensity of cold stress. There is also a faster restoration of normal skin temperatures once exposure to the cold environment has ceased. Other features of the acclimatized individual include an absence of reflex reactions from vessels in the mucous membranes of the nose and the upper respiratory tract, a reduction or even a total absence of the increments of heart rate and blood pressure initially seen in the cold, a reduction of shivering, and a stability of sensory chronaxie [Burton and Edholm, 1957; Curtis, 1985; Kojranovskij, 1966].

After the initial 'alarm reaction' (which persists for about 2 weeks), combinations of cold and hypoxia such as are encountered in continental Antarctica are often characterized by a drop in arterial pressures (particularly the systolic pressure) [Bobrov et al., 1979; Deryapa and Ryabina, 1977; Produtskij and Vorobyev, 1969]. In the latter part of a season's stay, there may be ballistocardiographic evidence of disturbances of myocardial contractility, with a greater asynchrony in contraction of the right and left ventricles [Davidenko, 1980]. A decline in physical working capacity has also been noted in residents at the 'Vostok' Research Station; this is apparent by the end of the first month of residence, or by the end of the polar night [Bobrov et al., 1979].

The decreased cardiac output, loss of orthostatic reactions, slowed atrioventricular conduction, extended electrical systole and altered systolic index may all be attributable to forced inactivity during the period of polar night. Stress-related extrasystoles and 'ischemic' changes in the electrocardiogram have also been described [Bobrov et al., 1979]. All of these physiological reactions are seen against the background of continued hypoxia.

Respiratory Reactions

The typical initial respiratory response to circumpolar residence is an increase in respiratory minute volume. This has been widely described as 'polar dyspnea', and is attributed by Soviet scientists to a combination of acidosis [Burton and Edholm, 1957] and decreased ionization of the air.

However, other factors contribute to the syndrome. The dyspnea is exacerbated by the need to move in heavy clothing and footwear on an uneven snow-covered surface, sometimes against a strong wind. There may also be a decreased efficiency of muscular work [Avtsyn et al., 1977] and a tendency to deterioration of bronchial conductance has been described in both smokers and people who have worked out of doors for long periods [Deryapa and Ryabina, 1977].

Psychomotor and Psychological Changes

A marked decrease in the comfort range of temperatures develops during wintering at the 'Molodezhnaya' Antarctic Research Station [Vasilevskij et al., 1978]. At the beginning of one expedition, the differential between unacceptable cold and warm temperatures was 15.3 °C, but by the end of the winter the range had narrowed to 9.3 °C. The threshold of cold sensitivity decreased by an average of 7.1 °C, whereas the threshold for warmth remained unchanged.

Polar explorers who reside in the coastal areas during the winter season show an overall decrease in the functional capacity of the central nervous system. This is manifested for example in a 28% increase in the overall error rate in self-monitoring of the electroencephalogram. Periods of high monitoring quality alternate with periods of poor performance. Error sign replacement is also performed more slowly [Vasilevskij et al., 1978]. By the end of the winter season, a further decline in cerebral function is seen.

Subjects who are healthy, psychically adapted and well motivated to adjust to the new conditions of high latitudes show a reduction in the duration of fast-wave sleep. On the other hand, psychically de-adapted subjects who have developed neurotic disturbances yet remain motivated to acclimatize show normal amounts of fast-wave sleep. There are differences between aboriginal and immigrant populations with respect to inter-hemispheric asymmetries of α-wave distribution, changes with functional loading of the brain, and specificity in spatial synchronization of potentials. In particular, the right cerebral hemisphere is more active in the high-latitude natives, and the left hemisphere is more active in the immigrants.

Conclusion

The conclusion from this brief historical survey of Soviet circumpolar research is that with the exception of a few limited survival studies on summer research teams, the Western literature offers no study of field personnel under conditions such as those encountered during the transpolar trek. There thus remains scope for a detailed report on the findings from a mission of this type.

4 Problem Formulation

*V.S. Koscheyev, V.K. Martens, T.S. Bashir-Zade,
A.A. Belyakov, G.V. Kypor, A.V. Visochanski,
M.A. Lartzev, A. Rode, M. Jetté, M. Malakhov*

Given the complex nature of the circumpolar environment and the equal complexity of human reactions, a systems approach is needed to survey individuals living and working in the north. Ideally, account should be taken of sex, age, length of residence in a given region, seasonal and diurnal rhythms, and the impact of atmospheric, geomagnetic and other factors. Specific and nonspecific adaptations should be distinguished in relation to qualitative and quantitative features of the exposure, and the biological characteristics of the individual, including nonspecific resistance and initial functional state [Baevskij, 1979; Geselevich, 1981].

It was not possible to control all of these variables on the ski-trek participants. Nevertheless, a first step was to define the initial functional status of the participants closely, describing their physiological and psychological potential and their nonspecific resistance to extreme environmental conditions. In addition to laboratory measurements of physiological data at rest and during exercise, a detailed biochemical evaluation, and thorough psychomotor and psychological testing, responses were evaluated during chamber simulations of the climatic conditions anticipated during the ski-trek. While the actual physical conditions were matched quite closely, the chamber experiments were unable to mimic the diverse range of other stressful experiences encountered during the traverse.

The study of acclimatization during the crossing sought to evaluate not only the extent of adjustments made by the body, but also the psychophysiological 'cost' of such adjustments, relating all data obtained to the observed features of the environment and the character of the required daily activity.

A final evaluation of the participants upon arrival in Canada was a further important feature of the study. At this stage in the research, the main task was to replicate initial measurements in order to assess the magnitude of any long-term functional changes that had developed during the course of the traverse. An attempt was also made to determine the speed of recovery processes, looking at both responses of the team as a whole and at interindividual differences.

Important new features of the overall mission were to examine interrelationships between physiological, psychophysiological and psychological function, and to relate each of these characteristics of the individual to personal success in coping with the requirements of the mission.

5 Personal Reflections and Medical Observations[1]
M. Buxton, J. Sproule, M. Jetté, A. Rode

There are few shores where one can stand, as Columbus did, and stare into the great unknown. Perhaps the Arctic Ocean is the last such place. Other oceans having been negotiated centuries ago, the inhospitable regions of the Arctic are the last to surrender their mystique to the map-makers. In the present era of technology, we know the depth, the ecology and the weather of the arctic, the limit of its expanse and what lies beyond. Yet without putting oneself adrift there and experiencing each of these variables, they lack real meaning.

It was in part the alluring mystery of the arctic, but also the relationship between the boundary nations that provided the impetus for the transpolar ski-trek. Dimitri Shparo is an accomplished Soviet arctic explorer for whom the magnetism of the north is more than a geophysical phenomenon. His 1979 expedition was the first to reach the North Pole on skis. The Polar Night Expedition of 1986, a 700-km trudge in total darkness, was a testament to the group's ability to adapt to adverse circumstances. As well, Shparo has a vision of what the arctic can become for its perimeter nations. As part of the new Soviet philosophy of Glasnost, the time was right for building a path across the Arctic Ocean. If Canada was also willing to participate, a ski-trek between the two nations could serve as a medium for unprecedented scientific, cultural and personal exchange.

Indeed, Canadians were ready to meet the challenge. The Department of External Affairs gave the Soviet leader their blessing, and 300 would-be explorers responded to the initial appeal for participants. Six were chosen to participate in training exercises and, of these, four were to be selected for the final trek. The Canadian contingent joined a multitude of Soviet candidates for a trek through the Tien Shan mountains of central Asia and then a week of winter camping at Iqualuit. After this initiation, four of the six recruits elected to withdraw, and with less than two months of preparation remaining, Chris Holloway and I (M. Buxton) joined the incumbents, L. Dexter and R. Weber.

[1] The medical examinations noted here were supported in part by the University of Ottawa Research Development Fund and the Health Services Centre of the University of Ottawa (Dr. D. Kilby). Josée Quenneville also contributed to the preparation of this chapter.

Our pool of expertise helped to complement that of the Soviet team. Laurie Dexter is an Anglican minister from Fort Smith, N.W.T., who has experience both as a radio operator and as an arctic traveller. Richard Weber was a member of the Steiger expedition, which reached the North Pole by dogsled without additional supplies in 1986. Chris Holloway, a longtime friend of Richard Weber, is a competitive skier and an excellent amateur photographer. I am a general practitioner who enjoys the cold.

Following a crescendo of preparations, we flew to Moscow on February 8, 1988. Our six Russian lessons scarcely enabled us to greet our hosts, but we had barely arrived when we found ourselves surrounded by reporters and scientists. Both were to become familiar expedition 'groupies' towards whom intense and varied emotions would be expressed over the next few months. The two groups took turns asking us questions and having us perform even odder tasks, all through the help of interpreters. We felt something of a novelty and, though our understanding of what was happening was limited, we nevertheless accepted our fate in the name of science and diplomacy.

Shortly after the expedition had been conceived, Soviet scientists had come to Canada in search of counterpart physiologists and psychologists to share in the research. The University of Ottawa agreed to host these studies, with input from Drs Andris Rode, Maurice Jetté and Roy Shephard.

The pretesting that was begun in Ottawa continued in Moscow. The expedition members' initial impressions of the program were twisted by the feeling that, but for the round of laboratory tests, they could be visiting the ballet, the circus or, at very least, skiing to prepare themselves for the trek. Any residual delusions about a skiing holiday certainly faded.

When the required data had been collected, we flew from Moscow to Dikson, a town of about the size and latitude of Iqualuit. Here we completed our preparations by testing ourselves and our equipment under arctic conditions, while Shparo made the final selection of the Soviet team.

It had been agreed by sponsors, government officials and team members that seven of the 12 Soviet candidates would be chosen. Equipment, food and overall strategy had been based on an 11-man team. It seemed strange to us that at this stage there were still 12 Soviet candidates. After a lengthy meeting, Shparo approached the Canadian contingent and proposed that the team be expanded to include an eighth Soviet participant. This posed some obvious logistic problems, but we had become good friends with our team mates and realized the importance of the project to each of them. Toli Fedyakov was a good man and we agreed to accomodate him.

5   Personal Reflections and Medical Observations

Further cuts were still required, and the air was tense as the final roll was called. Three of the original 12 were obviously crestfallen, but resigned themselves to serving as support staff. The fourth, Fiodor Konyikov, progressed from pleading to threatening, finally breaking a key piece of scientific equipment (our theodolyte). The logic behind what followed remains a mystery, but 13 people flew the next day to Seredny, the northern-most Soviet base, and then to our point of departure; Fiodor's tantrum had earned him a place on the team. Was this a harbinger of the leadership we were to expect?

In brief, the general strategy and daily routine of the ski-trek followed this pattern:

The plan called for the skiers to trek for 8–10 legs daily; each 1-hour leg included 50 min of travel followed by a 10-min rest.

The traverse was to be divided into five 12- to 14-day marches. At the end of each march the skiers would be resupplied by a parachute drop and then they would take a 2- to 3-day break for research, reorganization, and rest. The airdrops would provide food, fuel, equipment replacements, and mail. Three of the drops were planned on the Soviet side of the North Pole (March 15 and 30, April 14); there were to be a landing of supplies and dignitaries at the Pole (April 28) and one further drop on the Canadian side of the Pole (May 13).

On the trail, navigation was to be by observation with verification of the exact position each evening by a portable, 1-kg beacon linking with SARSAT, an international search and rescue satellite. The expedition was to maintain daily radio contact with a dedicated network of radio stations in the USSR, Canada, and Soviet and Canadian floating research stations on the Arctic ice cap.

A strange mixture of feelings surrounded our departure from the barren cape. The temperature was less than inviting, but the unexplained forces which had drawn us to this point were no less commanding. Moreover, we all felt a need to leave behind the confusion and conflict of the final preparations. Assembled on the tip of land was not only the ice party, but a hodge-podge of international journalists, scientists, organizers and well-wishers. Despite an unseasonably warm $-35\,°C$, there was an urgency of movement that would become very familiar in the days ahead.

The sun was a meager few degrees above the land behind us as we set off, and our shadows blended into the infinite blue ahead. Hoisting our fully laden packs onto our backs for the first time was a shock. Wearing 10 kg of clothing and carrying some 45 kg on our backs, we had difficulty in standing still. My first attempt at movement left me lying flat on my face staring out helplessly from beneath my burden. I got up that time, and perhaps 50 more times that day. There was some sadistic comfort in seeing one's teammates going through the same process. Somehow, we always managed to carry on, relatively unharmed.

The ice at Cape Arcticheski is unpredictable. Our reconnaisance plane had passed over the area a few days earlier, and had taken film footage of large, fairly smooth pans. Active currents and winds, we knew, could change the situation quickly, leaving us to face open water and rough ice. During the first week, we met both, repeatedly. It seemed that we would just get through one boulder field and we would be met by a fresh crack. These invariably seemed to lie perpendicular to our course, forcing us miles to left or right in search of a crossing point.

And the ice was alive. As one huge floating pan shifted relative to the next, it would heave the edges into piles of debris 6 m high. The sound was eerie, and haunted our sleep as the world agitated around us. At first, it was like the noise of engines, smooth and regular in the distance, but as it came closer and we saw the actual ridges forming, it resembled machine-gun fire, and we reacted to it in the same way.

The most horrifying moment of the trip took place as we fled from a campsite that nature had slated for demolition. A deafening sound awoke us, and we wasted no time in breaking camp. We sought an escape route, only to find the perimeter of our ice-flow surrounded by water. The only point of contact with another body of ice was being grated into car-sized boulders, folding under and being thrown up on the edges of the pan as they moved. One by one, the skiers made their way across. When my turn came, my worst fears were realized. I slipped and fell into the churning chasm. The images of death by simultaneous crushing, drowning and freezing rushed through my mind. I grovelled for an eternal second in purgatory before finding myself inexplicably on solid ice once more. I had been plucked out of the water by a comrade from ahead and a buddy from behind. From that point onwards, national identities faded and we began to appear as a single organism, snaking in unison across the ice on our 13 pairs of legs.

Our day started at 6 a.m., as the member on kitchen duty beckoned us to breakfast. One by one, the groggy coccoons would metamorphose into dishevelled snow nymphs and consume with insect-like voracity a liter of the grey, oily slop that sustained us. A cup of weak instant coffee was taken to maintain hydration and induce peristalsis, and we then moved out into the elements with our ration of toilet paper to perform the least pleasant task of the day.

Breaking camp was particularly tedious in the cold, and we were relieved to be up on our skis for the first march of the day, generating an increased amount of body heat. After 50 min of skiing, we rested for the carefully timed remainder of an hour before continuing. Any longer rest led to dangerous cooling and numbness of the extremities. During the breaks, we nibbled continuously on chocolate, dehydrated fruit and pork fat. Nine such cycles were repeated in the course of an average day.

At the end of the day, we selected a flat piece of old, solid ice, and arranged our skis in a circle, with the tails stuck into the snow. An aluminum frame united the tips, and a nylon cover was drawn over the top to create a light efficient tent which broke even the most severe wind.

Each of us had specific scientific or practical tasks to perform while the cook of the day was melting snow for the evening meal. After brushing the frozen perspiration vapor from our clothing, we climbed inside the tent, and hung our ice-caked hats, boots and masks from the ceiling. We then arranged ourselves radially in our sleeping bags, with our feet in the middle. It was at this point each day that the greatest conflict developed, as latecomers into the crowded space jostled for position.

Every night, another 13 liters of grey oily slop was served, differing only slightly in composition from the morning meal. It consisted of buckwheat groats, dehydrated milk and butter, freeze-dried hamburger and pemmican. In the morning, muesli was substituted for the pemmican. Each serving delivered about 12.5 MJ of food energy. Two servings per day, with snacks taken on the trail, provided over 25 MJ.

The high-fat content took its toll on our digestive systems. Within the first week, almost all of us had developed diarrhea. The problem was unbearable in the extreme cold, and we were most relieved when the first supply drop included the lipase tablets that we had requested. Our digestive efficiency then improved, and we no longer had to squat along the trail at half-hour intervals.

The tent was a shelter from the wind, but it provided few other comforts. The inside temperature was usually about 6 °C warmer than outside. When the stoves were blazing under the cookpots, it would creep above freezing just long enough for the condensed vapor to melt and flow from the ceiling onto our sleeping bags. The warmth and moisture that it generated was more detrimental than helpful.

On the coldest nights, -50 °C, we lay awake in our saturated sleeping bags, shivering and afraid to lose the vigil we kept over the precious sensation in our hands and feet.

My Soviet counterpart, Dr. Misha Malakhov, and I were presented with a number of medical problems during the first weeks of the trek, particularly blisters, frostbite, acute musculoskeletal injuries and abdominal pain from our fat-induced enteritis. Remarkably, we were prepared for all of these needs, and not one day of travel was lost for health reasons, although some members of the expedition were in danger of losing some of their smaller appendages from cold exposure.

At the end of each 2 weeks of travel, with our supplies dwindling, the resupply plane was called. Within 12 h, it would appear from out of the southern sky to drop our food, fuel and replacement equipment. The 2 days that followed were spent in repacking, snoozing and gorging on food that had been specially prepared for the rest days. Some of it was home baking from wives and mothers; all of it was eaten. For a few brief minutes, we even got warm ... and clean. Extra gasoline was burned on the stoves, and by draping the parachutes over the tent we created a makeshift sauna. After luxuriating with a cup of bathwater, we took advantage of the exposed flesh to collect physiological data – skinfolds, weights, and blood samples – for later analysis.

It was difficult to remobilize after the rest, especially with the packs restored to their original weights. The skiing itself was not very exciting in the first half of the trip. Every step was an effort. There was no glide on snow that seemed the consistency of sand, and our restrictive clothing and single-file pattern made communication difficult, even amongst the Canadians. The experience was somewhat akin to sensory deprivation, and the effects upon the psyche were strange.

I spent my mental energy on thinking about when I would next experience pleasure, and in what forms it would come: the next rest break, the next meal, the next letter from home, the arrival there. The emotions underneath all that insulation became quite intense at times, and I found myself alternating between sobs of melancholy and hysterical laughter. Sometimes I would ruminate on verses of songs, or make up nonsense verses that would act as my mantra until I skied blindly into the body in front of me at the hourly resting stop. And always there was a faint fear of death, surfacing subtly into consciousness when it was permitted. Combined with the chances to assess personal feelings without outside interference, it was enough to inspire me to matrimony, and on bended knee in front of the radio at the end of an exhausting day, I proposed marriage to Nancy, who is now my wife.

My impressions of the first half of the trip were warped by pain. A badly frostbitten foot kept me distracted and limping. Misha advocated polypharmaceutic treatment with which I can find no fault in light of the eventual positive outcome. Even more helpful, though, was the distribution of some weight from my pack among the other skiers, a practice which kept the group together despite other ailments. The quality of life each day was very dependent upon the weather. It was hard to be cheerful when skiing against 70-km/h headwinds. In contrast, tranquil sunny days seemed ecstasy. Though no heat was felt from the sun, its presence was reassuring, and the highlights it gave to the bizarre ice-forms created some of nature's finest works of art.

March 28, less than a month into the trek, marked the beginning of 'polar day'. Uri and I stood outside the tent and clapped as the sun shrank to a thin line of orange on the horizon, then began its glorious resurgence. We had shared the role of 'worst victim of frostbite' in the preceding weeks, and the new era of light and potential warmth had particular meaning for us. Uri was a mathematics professor and the senior member of our group at the age of 52 years. With frost having eaten the best half of his nose, his appearance seemed that of a crashed fighter pilot. He was not concerned, he said, because he was in charge of geomagnetic programs. The nose was a medical problem, and therefore the concern of the doctors. As our personalities thus revealed themselves, I began to realize what a pack of eccentrics we were.

The politics of decision-making on the ice was certainly more democratic than it had been in Dikson. There were few choices to make, but when opinions differed about the best way to cross a lead, or when to make camp, democracy reigned. Shparo was far from totalitarian, and the other Soviets found the concept of voting to be entertaining.

Relations between individuals at times became strained, but the conflicts were seldom of an international nature. The group was large enough for adversarial personalities to avoid each other. Shparo was amazed at our ability to maintain morale. On previous expeditions, severe conflicts had nearly destroyed his group and resulted in failure. Even Fiodor, whose actions had alienated him initially, had reintegrated with the team and become a very solid travelling companion.

The days went by slowly at first, but the weight of the journey ahead diminished with each day, each stage, each step. The whole idea became more tangible as we received our coordinates and plotted our progress on the map each evening. Our confidence increased, and 55 days after leaving the desolate Siberian shore, we reached the North Pole. It was hardly as I would have expected it. Visually, it was no different from any other region of frozen ocean, except for a small inflatable village and 180 people assembled there to greet us. This was the biggest party the Pole had ever seen, including reporters, filmmakers, and such distinguished guests as: Marcel Masse, Canadian Minister of Energy, Mines and Resources; Albert Reichman, Chairman of Olympia and York Developments; Alexis Axionov, pioneer Soviet cosmonaut, and Uri Israel, Soviet Minister of Hydrometeorology.

We were a motley, but very happy crew as we reestablished links with the southern world. We had achieved our first goal, and had survived the longest and most arduous portion of the trek. The reception we received was perceived as the strongest possible vote of confidence from the outside world. The 13 frost-scarred faces circulated within the crowd. Portly politicians danced arm-in-arm amid a cacophony of fireworks and popping champagne corks. Speeches were made, interviews given, and gifts exchanged. It felt as though the Canadian shore was just over the horizon.

But as quickly as we had fallen upon this event, it faded, like an abruptly ending dream. News of worsening weather drove the crowds back to their helicopters, and they were gone. We were alone again. The storm from which the crowd was fleeing brought warm winds and whiteout conditions. Spring was on the way, and it was urgent to press forward. A huge crack formed through our ceremonial grounds, and we learned later that the Soviet drifting station which had acted as our transportation and communication relay center had broken up, and part of it had slid through the ice.

The homeward portion of the trip was fraught with its own problems. Turbulence from the merging of two ocean currents on the Canadian side of the Pole created endless miles of rubble. Warmer temperatures produced new cracks, and areas of open water failed to refreeze, making detours longer, and opportunities for unexpected swims more frequent. Increasing humidity brought fog that destroyed the visual contrasts. Ridges and holes blended into an ubiquitous white, making formidable traps for the lead skiers.

On the positive side, life was no longer a struggle against the penetrating cold. We now had stamina, experience and confidence, with a less easily defined group mania that spurred us on with boundless energy. The concepts of day and night lost their relevance, with the sun circling overhead, and we struggled to maintain our established circadian rhythms. Skiing became more like a sport and less like a penance, and we spent more and more time on the trail. The clear days were now beyond ecstasy. Ice formations captured the light in ever more fantastic ways as their cloaks of snow sublimed away.

Satellite photographs taken in the final weeks were a perpetual source of controversy and dismay. The image of a huge lead averaging 40 km in width appeared repeatedly. An obstacle of this magnitude was something none had reckoned with. Our one-man boat could hardly ferry the group across, and even if a larger boat were to become available, the danger would be too great. Rather than consider the likelihood of failure, we tried to put the issue from our minds. We knew that the distance was getting smaller day by day, but the suspense leading up to the crossing of each new ridge became unbearable.

The last day of May was white, like many previous days. Visibility was poor, and the endless motion of my own ski tips was the only sight of interest. We moved like zombies, with our thoughts elsewhere, far away. Few of us noted the faint patches of blue that appeared ahead at first, but an excited cry of disbelief from Valodia at the head of the line quickly focussed our attention. I stood transfixed, as the mountains of North Ellesmere emerged from the haze. The incredible natural beauty of these seldom seen peaks was magnified a hundredfold by the 3-month quest.

The remaining 30 km slid under our feet virtually unnoticed. Somehow the great lead, the impassable obstacle, had disappeared. Whether it froze,

5 Personal Reflections and Medical Observations

blew closed, or was a figment of the satellite's imagination will remain a mystery. Our last day of travel was the most spectacular. The temperature was $-4\,°C$, but the brilliant sunshine felt much warmer. The pack-ice in the vicinity of Ward-Hunt Island, where we finally made contact with Canada, contained the largest and most beautiful formations of all. One would have expected us to be exhausted. True, we took turns rolling at length in the first patch of gravel we found, but we were soon on our feet, exploring the island's contours and communing with the tiny life-forms that were celebrating the arrival of spring.

We made contact with our base in Resolute Bay, and some time thereafter a Twin Otter aircraft appeared to collect us and take us home. Despite the anguish, the pain, the boredom, the conflict and everything else we had come to hate about the Arctic, a big part of me refused to board the plane. The ice had become home – no longer a desolate wasteland separating the two largest countries on earth, but a bridge uniting them. The pleasures I had envisaged in the early days of the trip were now realized – the reunions, the accolades, the warmth, marriage and induction into the order of Friendship of Nations all exceeded the capacity of my icebound imagination. But the most valuable of the trek's dividends was intimacy – with the people who live across the ice, and with the exotic expanse which joins us, the mysterious North.

Medical Observations

From the logistic and medical viewpoint, the expedition was well organized and maintained its schedule. No time was lost from accidents or injury, and despite some setbacks, there were no life-threatening incidents during the entire mission. The success of the trip seems attributable to a number of factors, including the experience of the skiers (10 of the 13 team members had previous experience of polar expeditions), good planning and organization, good logistical and medical support and excellent communications, individual motivation and determination, and a team spirit that eventually developed as the expedition progressed.

Immediately following the trek, medical examinations were conducted at the Health Services Unit, University of Ottawa. All members of the expedition were examined by the same physician within 48 h of reaching the Canadian Arctic coast. The examination included a thorough history and clinical examination, and electrocardiogram, blood and urine analysis, pulmonary function tests and chest radiographs.

In general, all members of the team seemed in good physical health and in good spirits. Every skier had suffered some cold injury, but most lesions had resolved by the time of the medical examination. Some subjects showed residual healing cold injury of the fingers and toes, primarily erythema and scaling skin, and there were two cases of severe frostbite. One man had severe frostbite of the end of his nose, with loss of tissue and formation of an eschar, the damage being permanent. Another team member suffered frostbite of the end of the great toe, resulting in eschar formation and loss of sensation.

The other major problem during the trek was prolonged exposure to wind and solar radiation. Despite the use of conventional chemical sun blocks, all team members suffered some degree of skin damage. As anticipated, the problem was more severe in those with a fair complexion. The radiation problem increased as the hours of sunlight became longer during the latter stages of the expedition; it was hypothesized that the magnitude of skin damage was exacerbated by the thinner ozone layer in the arctic atmosphere. Most of the skiers also reported eye irritation, both from the intense direct sunlight and from light reflected by snow, ice and water.

Other problems during the expedition included mechanical trauma to the feet and toes, and an irritant rash thought to be provoked by the undergarments. The latter was aggravated by kneeling in the snow while setting up camp and doing the routine housekeeping chores in the tent morning and evening.

Most of the skiers reported gastrointestinal discomfort, primarily in the form of cramps and diarrhea. This occurred mainly during the early stages of the trek, and was thought due to the high-fat diet that was being eaten at the time. The problem resolved after several days.

Muscular aches and pains were common throughout the trek. Most of the team reported low back pain, which was attributed to the heavy backpacks. There were complaints of typical overuse injuries such as tibialis anterior tendinitis. The heavy load carried in the backpacks was also blamed for transient numbness to the arms and hands.

The electrocardiograms, blood (table 5.1) and urine analyses and chest radiographs were all normal. We were surprised to find that three of the Soviet participants had a history of mild hypertension, which was also evident on clinical examination. All of the skiers were nonsmokers. Pulmonary function tests nevertheless revealed three individuals with mild restrictive lung disease; these conditions were judged to be chronic, since there had been no complaints of respiratory problems during the trek.

There were some interesting differences between the Canadian and the Soviet team members. On average, the Soviet participants were older and heavier, and had a higher resting heart rate, a higher resting blood pressure, and a higher peak exercise blood pressure. The Soviet team also had a higher average percent body fat, and overall they lost weight during the expedition. In general, they could be characterized as robust,

Table 5.1. Blood analysis for the Canadians (n = 3) before and after the trek (mean values)

Variables	Before trek	After trek
Glucose, mmol/l	5.1	4.8
Urea, mmol/l	6.5	5.8
Total Protein, g/l	71*	66*
Albumin, g/l	49*	46*
Cholesterol, mmol/l	5.44	6.32
Triglyceride, mmol/l	0.60	0.96
Calcium, mmol/l	2.25	2.19
Sodium, mmol/l	136	139
CO_2, mEq/l	23*	28*
Creatinine, μmol/l	82	81
T_3 uptake, nmol/m	0.42	0.42
T_4, nmol/l	98	93
Free T_4 index	41	38

* Significant difference at $p < 0.05$.

hardworking and adventurous individuals who were used to accepting hardship. This can be illustrated by the Soviet participant with the severely frostbitten nose. When asked about the resultant pain, he replied that it was not his problem, it was the team physician's problem.

The Canadian participants, on the other hand, were younger, lighter, and of slimmer build. They had lower resting heart rates, lower resting blood pressures, and lower peak exercise pressures. They also had a lower percentage of body fat, and actually gained weight during the expedition, while their maximal oxygen intake dropped by about 20%; it was hypothesized that training prior to the trip had been more vigorous than the activities encountered during the expedition. Overall, the Canadians had the characteristics of slim, long-distance endurance athletes, and appeared to be very attuned to their bodies. They were able to recall readily and to describe clearly numerous minor injuries such as tendinitis, rashes and sore muscles that they developed during the trek.

It was concluded from examining and questioning the team members that they had not been severely stressed from the viewpoint of cardiorespiratory endurance. Rather, the experience had been a battle against the elements in a very unfriendly environment. By far the most important medical problem was skin damage from the combined effects of sun, wind and cold. The chemical blocking agents that were available offered protection against ultraviolet rays A and B, but did not protect against ultraviolet

C rays; the last are normally screened out by the ozone layer, but screening is insufficient at high latitudes. Opaque physical sun blocks that include zinc oxides and other pastes applied directly to the skin should prevent all solar radiation from reaching the skin, and would also provide some physical protection against wind and cold. Future expeditions should give more thought to the choice of appropriate blocking agents.

6 Cardiovascular Function and Autonomic Regulation[1]

V.S. Koscheyev, V.K. Martens, T.S. Bashir-Zade,
A.A. Belyakov, G.V. Kypor, A.V. Visochanski, A. Rode,
M. Jetté, M. Malakhov

Methodology

Soviet data on cardiovascular function comprised a full evaluation of the resting electrocardiogram and the resting ballistocardiogram. The Soviet scientists were interested to interpret their findings in the broad spectrum of adaptation to stress, whether physical or psychological. In their view, a person who was well-adapted would show manifestations of parasympathetic tonus such as sinus arrhythmia and a slow resting heart rate, while a person who was poorly adapted would show a dominance of sympathetic activity, with a high resting blood pressure, suppression of sinus arrhythmia and a high resting heart rate.

The electrocardiographic data included measurements of resting heart rate, electrical axis, P and T wave amplitudes, P wave duration, PQ interval, MacRause index (P/sPQ), PQ Interval, QT interval, systolic index, QT/TQ ratio, Bazette index ($QT/\sqrt{RR}$), activation time of the right ventricle, and activation time of the left ventricle.

Such detailed analysis of the resting record is uncommon in Western laboratories, and it may thus be useful to comment on some of the measurements. The normal electrical axis ranges quite widely from $-38°$ to $+90°$, a right axis deviation being seen in a thin individual with a vertically placed heart, and left axis deviation (readings of less than $-30°$) being seen with both obesity and marked left ventricular hypertrophy. An increase in amplitude or duration of the P wave is quite common in endurance athletes [Morganroth and Maron, 1977]. The McRause index is not commonly used in the West, but is another method of expressing this relationship. The duration of the various portions of the ECG waveform is determined by the rate of conduction of electrical impulses through the myocardium. In general, such conduction is prolonged by parasympathetic tone, and it is thus common practice to 'correct' conduction times by reference to heart rate, as in the Bazette index of QT times. A partial right bundle-block is also a common finding in well-trained subjects with a high level of parasympathetic tonus.

---

[1] Freely adapted from the Russian translation by R.J.S.

Table 6.1. Initial ECG characteristics of the 13 participants

Participant	Heart rate bpm	Cardiac axis	Electrical axis degrees	P wave (s PQ) s	PQ interval s	MacRause index
D.I. Shparo	53	semi-horizontal	25	0.05	0.16	2.20 F.F.
Konyukhov	80	vertical	80	0.07	0.16	1.98 V.I.
Shishkarev	54	normal	40	0.08	0.18	1.25 M.G.
Malakhov	53	semi-vertical	70	0.08	0.15	1.14 A.V.
Melnilov	60	horizontal	0	0.10	0.22	1.20 A.A.
Belyaev	77	vertical	80	0.06	0.17	1.83 A.P.
Fedyakov	55	vertical	90	0.11	0.21	0.91
Yu.I. Khmelevsky	57	normal	30	0.06	0.17	1.80 V.P.
Ledenev	73	semi-vertical	70	0.05	0.15	2.00 M.
Buxton	41	vertical	85	0.08	0.18	1.25 L.
Dexter	55	horizontal	20	0.08	0.18	1.25 R.
Weber	56	vertical	80	0.02	0.10	4.00 K.
Holloway	75	normal	40	0.09	0.20	1.20
Mean ± SD	59.0 ± 2.8			0.07 ± 0.01	0.17 ± 0.01	1.69 ± 0.22

MacRause index = P/sPQ; Bazette index = $QT/\sqrt{RR}$; Activation time of right ventricle = $[P-(iQT/RR)] \times 100\%$.

Baseline Data

The initial resting electrocardiograms for the 13 participants in the ski-trek show values within accepted physiological limits (table 6.1). Eleven of the 13 participants had a normal sinus rhythm, but there was a marked sinus arrhythmia in two members of the team. The electrical axis was intermediate in three subjects, semi-horizontal in one, horizontal in two, semi-vertical in one and vertical in five participants. Lengthening of atrial systole, which the Soviet cardiologists interpret as evidence of an increased loading of the right side of the heart, was seen in L.D., K.H., A.P.F., V.I.S. and M.B. Three subjects showed a clinically insignificant slowing of A-V conduction (R.W., V.P.L. and M.G.M), and in D.I.S there was an incomplete blockade of bundle of His. Given the absence of either symptoms or marked deviations of electrical axis, all of these findings were accepted as falling within normal limits. Three participants showed functional 'disturbances' that the authors would attribute to the slow heart rate and high level of parasympathetic tone in these particular individuals: slowing of atrial conduction (A.P.F.), atrioventricular conduction (A.V.M) and intraventricular conduction (M.G.M.).

QRS interval s	QT interval s	Systolic index %	QT/QR	Bazette index	Activation time, s	
					right ventricle	left ventricle
0.10	0.41	36	0.59	0.38	0.01	0.04
0.10	0.40	43	0.71	0.41	0.02	0.04
0.08	0.37	33	0.49	0.35	0.02	0.02
0.12	0.41	36	0.60	0.38	0.03	0.04
0.08	0.40	40	0.66	0.40	0.03	0.04
0.08	0.35	45	0.81	0.40	0.03	0.04
0.07	0.40	36	0.57	0.38	0.02	0.04
0.09	0.39	37	0.57	0.38	0.02	0.03
0.08	0.36	44	0.81	0.40	0.02	0.04
0.09	0.42	29	0.43	0.35	0.02	0.04
0.08	0.40	36	0.56	0.38	0.02	0.04
0.10	0.42	39	0.45	0.40	0.03	0.04
0.10	0.35	44	0.76	0.39	0.02	0.04
0.09 ± 0.00	0.39 ± 0.01	38.0 ± 1.3	0.61 ± 0.03	0.38 ± 0.00	0.02 ± 0.00	0.04 ± 0.00

The satisfactory initial cardiovascular status was confirmed by ballistocardiographic observations made in Moscow immediately before departure on the trek (table 6.2). The resting heart rate, arterial pressure, pulse pressure, mean driving pressure, and estimates of resting stroke volume and cardiac output all corresponded closely with anticipated laboratory norms (although the absolute values for cardiac output seem somewhat low). The negative score for Kerdot's autonomic index was considered further evidence that parasympathetic drive predominated in most subjects. This index, again not common in the West, evaluates sinus arrhythmia in terms of differences between the arithmetic mean and the mode of heart rate in individual subjects. A negative index implies an arrhythmia, while a small size of the difference between mean and mode is interpreted by the Soviet cardiologists as evidence of a state of complete physical and psychic rest.

Two subjects (A.V.M. and A.P.F.) showed hypertension, with a decrease in resting stroke volume and cardiac output. This is interpreted by the Soviet cardiologists as evidence of autonomic imbalance, with activation of both parasympathetic and sympathetic centers arising from overstrain of neural mechanisms, an irradiation of impulses to lower control

Table 6.2. Hemodynamic data obtained on the participants before departure (Moscow)

Participant	Heart rate bpm	Arterial pressure, mm Hg		Pulse pressure mm Hg	Mean driving pressure mm Hg	Stroke volume ml/beat	Cardiac output ml/min	Kerdot's autonomic index %
		systolic	diastolic					
D.I. Shparo	72	115	80	35	91	48	3,449	−11
F.F. Konyukhov	86	120	80	40	93	47	4,029	6
V.I. Shishkarev	69	115	70	45	85	62	4,271	−1
M.G. Malakhov	56	130	60	70	83	72	4,032	27
A.V. Melnikov	60	155	100	55	118	39	2,322	−67
A.A. Belyaev	80	140	90	50	107	48	3,856	−11
A.P. Fedyakov	55	155	100	55	118	39	2,129	−82
Yu.I. Khmelevsky	80	135	70	65	92	63	3,032	12
V.P. Ledenev	89	115	70	45	85	50	4,441	21
M. Buxton	66	100	70	30	80	53	3,472	−6
L. Dexter	65	125	70	55	83	68	4,427	−8
R. Weber	73	110	80	30	90	50	3,665	−9
Mean ± SD	71 ± 5	126 ± 12	78 ± 24	48 ± 23	94 ± 17	63 ± 14	3,760 ± 232	−13 ± 40

centers, and inadequate 'adaptation'. Plainly, both of the subjects concerned had low resting heart rates. From a Western perspective, such high diastolic pressures might have been regarded as a contraindication to participation in such a strenuous undertaking, but both subjects completed the trek without symptoms or clinical incident.

Both F.F.K. and V.P.L. showed an above-average level of sympathetic activity, as judged by the short mean and mode of cardiac cycle length (table 6.3). The second slow wave was also longer than the respiratory wave, suggesting an increased activity of subcortical centers, and a lack of normal sinus arrhythmia was underlined by low values for the variance of heart rate. The values for mode, amplitude and arbitrary stress index suggested to the Soviet cardiologists that there was an activation of both humoral and nervous mechanisms in these individuals, with a predominance of sympathetic tone. This was further suggested to be a defensive-adaptive type of stress in individuals who had a capacity to adapt. In Y.I.K., the short cardiac cycle duration was not matched by amplitude and stress index data, the latter being close to the population averages. Furthermore, the periodicity of the respiratory waves was much greater than that of the slow waves; the Soviet cardiologists interpreted these findings as evidence of deregulation, or an unsatisfactory adaptation to the overall stress.

In A.A.B., M.G.M. and K.H., most indices were close to population values (table 6.3). They also showed a longer period for respiratory than for slow waves, with Kerdot's index and heart rate suggesting a parasympathetic dominance. In general, this picture suggested a healthy adaptation, although in K.H. the high values of amplitude and stress index suggested some stressing of the regulatory mechanisms. In M.B., L.D. and V.I.S. the findings were generally indicative of adequate adaptation, although the increased slow wave duration suggested some physical or mental fatigue. The low level of sympathetic function and adequate adaptation in D.S. and R.W. were shown by long cardiac cycle lengths, respiratory cycle lengths that were much longer than slow waves, and sinus arrhythmia leading to a large variance of cycle length and a low stress index.

In summary, the Soviet cardiologists concluded from the initial examinations that three subjects showed inadequate adaptation (Y.I.M., A.V.M. and A.P.F.), and three showed clinically significant ECG changes (A.P.F., M.G.M. and A.V.M., two of these latter three being the hypertensive individuals previously noted).

Observations during the Expedition

During the expedition, the only electrocardiographic data of sufficient reliability were obtained at the fifth drop, after the subjects had completed about two thirds of their journey (table 6.4). When compared to the initial status, there was significant lengthening of the average cardiac cycle length ($p < 0.05$). Western scientists would probably interpret this as a training response, although the Soviet cardiologists noted small increases in amplitude and stress indices, with a decrease in variance as evidence of a growing stress upon regulatory mechanisms, with a narrowing of potential adaptive reactions, and signs of a mismatching of nervous and humoral influences upon the heart.

An increase in the slow-wave constituents of the cardiac cycle was considered evidence of an activation of sympathetic subcortical centers, operating against a background of enhanced parasympathetic activity, and a centralization of control processes. However, it was also noted that there were substantial interindividual differences in slow wave and respiratory wave changes, suggesting that in some subjects subcortical activity was enhanced, and in others it was depressed.

Three groups apparently differed in their pattern of adaptation to the expedition. Adequate adaptation was shown by M.B., L.D., R.W., K.H., V.P.L., V.L.S., D.I.S. and F.F.K., although the last two of these eight individuals showed some changes that could be interpreted as evidence of physical or mental fatigue. Two subjects (M.G.M. and A.A.B.) showed inadequate adaptation, and three people (A.P.F., A.V.M. and Y.I.K., the

Table 6.3. Cardiac rhythm indices of the participants before beginning the trek (Dikson)

Participant	Cardiac cycle, ms		Coefficient of variation %	Mode, cardiac cycle ms
	mean	SD		
D.I. Shparo	1,036	57	5.5	1,010
F.F. Konyukhov	678	19	2.8	700
V.I. Shishkarev	852	38	4.5	855
M.G. Malakhov	903	84	9.3	940
A.V. Melnikov	818	41	5.0	825
A.A. Belyaev	932	45	4.8	925
A.P. Fedyakov	899	45	5.0	890
Yu.I. Khmelevsky	720	43	6.1	705
V.P. Ledenev	700	29	4.1	720
M. Buxton	821	61	7.4	795
L. Dexter	829	98	11.8	810
R. Weber	1,034	76	7.1	1,060
K. Holloway	847	30	3.5	845
Mean ± SD	836 ± 102	49 ± 23	5.8 ± 2.5	835 ± 97

Table 6.4. Cardiac rhythm indices of the participants at the 'fifth drop' with two thirds of the journey completed

Participant	Cardiac cycle, ms		Coefficient of variation %	Mode, cardiac cycle ms
	mean	SD		
D.I. Shparo	1,003	72	7.2	1,025
F.F. Konyukhov	913	39	4.2	865
V.I. Shshkarev	880	99	11.3	945
M.G. Malakhov	965	27	2.8	965
A.V. Melnikov	1,056	32	3.0	1,025
A.A. Belyaev	958	38	3.0	975
A.P. Fedyakov	1,030	19	1.8	1,025
Yu.I. Khmelensky	907	27	3.0	895
V.P. Ledenev	876	117	13.3	765
M. Buxton	955	111	11.6	805
L. Dexter	843	53	6.3	779
R. Weber	962	59	6.2	975
K. Halloway	823	24	2.9	855
Mean ± SD	941 ± 71	46 ± 25	5.4 ± 3.7	913 ± 96

Amplitude about mode %	Range of variation ms	Respiratory wave s	Slow waves, s		Stress index arbitrary units
			SW-1	SW-2	
32	265	67.9	3.9	28.1	60
60	105	13.4	36.7	49.9	408
44	160	12.8	14.0	73.2	161
23	360	32.7	51.9	15.4	34
42	180	40.2	24.1	35.8	141
41	245	38.0	29.2	32.8	90
48	260	49.2	15.0	35.8	104
50	200	76.4	10.0	13.5	177
58	115	35.4	24.4	40.2	350
43	290	12.9	38.2	48.6	93
32	400	12.6	51.3	36.1	49
14	452	70.2	2.8	29.0	14
57	160	51.9	10.6	37.5	211
44 ± 11	228 ± 92	40.7 ± 21.5	22.2 ± 16.0	36.1 ± 18.0	157 ± 1

Amplitude about mode %	Range of variation s	Respiratory wave s	Slow waves, s		Stress index arbitrary units
			SW-1	SW-2	
38	320	9.1	13.5	77.4	58
40	160	21.0	18.2	60.9	145
51	320	4.9	13.4	81.6	84
73	160	46.1	12.0	42.0	236
55	200	65.5	20.3	14.2	136
47	102	62.8	11.4	25.8	236
92	80	41.7	25.6	32.8	561
51	160	17.2	24.6	58.3	178
27	380	17.9	17.2	65.0	46
22	380	46.7	23.5	29.7	36
29	256	51.7	30.8	17.5	73
34	360	66.1	18.7	15.1	48
50	100	47.4	43.3	9.3	292
47 ± 20	111 ± 113	33.7 ± 17.2	20.9 ± 14.7	40.7 ± 21.0	170 ± 150

Table 6.5. ECG indices measured after completion of the expedition (Ottawa)

Participant	Heart rate bpm	Heart position	Electrical axis degrees	P wave (sPQ) s	PQ interval s
D.I. Shparo	63	deviation to the left	30	0.040	0.130
F.F. Konyukhov	74	semivertical	70	0.030	0.150
V.I. Shishkarev	73	normal	50	0.070	0.180
M.G. Malakhov	63	normal	60	0.040	0.140
A.V. Melnikov	61	normal	40	0.080	0.200
A.A. Belyaev	77	vertical	70	0.30	0.110
A.P. Fedyakov	68	semivertical	70	0.070	0.190
Yu. I. Khmelevsky	64	semivertical	70	0.050	0.160
V.P. Ledenev	73	vertical	80	0.040	0.140
M. Buxton	64	vertical	80	0.040	0.160
L. Dexter	58	horizontal	10	0.030	0.150
K. Holloway	61	vertical	80	0.040	0.120
Mean ± SD	65.80 ± 1.44			0.050 ± 0.007	0.160 ± 0.09

Reliable differences from the initial status ($p < 0.05$): MacRause index. Bazette index, Fogelson index. $P = (iQT/RR) \times 100\%$.

last two having unsatisfactory initial records) showed both inadequate adaptation and signs of deregulation.

Observations on Completion of the Expedition

Electrocardiograms obtained in Ottawa immediately after completion of the ski-trek showed a normal sinus rhythm in 11 subjects, and a marked sinus arrhythmia in two individuals. Changes in the electrical axis of the heart were neither systematic nor important (table 6.5). In six subjects, the axes were substantially unaltered, in four there was an insignificant change to the left, and in three a shift to the right.

However, there were systematic changes in other features of the ECG ($p < 0.05$), including a speeding of atrioventricular conduction, a slowing of right ventricular activation and an increase in the MacRause index, suggesting an overloading of the right side of the heart. The broadening of the QRS complex and the greater QT/TQ ratio speak to changes in the processes of myocardial activation and repolarization.

MacRause index	QRS s	QT interval s	P, %	QT/TQ	Bazette index	Duration of right ventricular activation s
2.25	0.120	0.40	42.0	0.71	0.410	0.040
4.00	0.080	0.37	46.0	0.84	0.410	0.030
1.57	0.100	0.38	46.0	0.86	0.420	0.020
2.50	0.120	0.39	41.0	0.68	0.400	0.040
1.50	0.090	0.40	41.0	0.69	0.400	0.030
2.00	0.110	0.39	41.0	0.71	0.400	0.040
1.71	0.080	0.40	45.0	0.83	0.430	0.040
2.20	0.080	0.36	38.0	0.63	0.370	0.040
2.50	0.090	0.37	45.0	0.82	0.410	0.040
3.00	0.110	0.40	39.0	0.64	0.400	0.030
4.00	0.090	0.27	39.0	0.65	0.380	0.030
2.00	0.120	0.40	31.0	0.69	0.400	0.030
2.41 ± 0.25	0.100 ± 0.005	0.38 ± 0.04	42.1 ± 0.8	0.73 ± 0.02	0.400 ± 0.004	0.034 ± 0.002

A number of subjects showed functional changes, including a retardation of intraventricular conduction (M.B., K.H., R.W., D.I.S., M.G.M., A.A.B.), and high values of the MacRause index, combined with prolonged right ventricular activation (L.D., M.B., D.I.S., Y.I.K., V.P.L., M.G.M., A.A.B.). These findings again suggest an overloading of the right heart.

Ballistocardiography (table 6.6) showed a decrease in resting heart rate ($p < 0.05$) relative to the initial data, although values at this stage were apparently higher than at the 'fifth drop' during the expedition. No significant changes in resting arterial pressure, stroke volume or cardiac output were seen.

A continued stressing of regulatory mechanisms was suggested by an increase in amplitude and stress indices, with a decrease in the variance of the resting heart rate (table 6.7) relative to both initial and 'fifth drop' data.

Looking at the results for the individual subjects, the adverse findings in A.V.M. and A.P.F. noted at the initial examination (hypertension, low stroke volume, low cardiac output and substantial negative Kerdot index) persisted. V.P.L. showed similar but less marked findings, indicating a stressing of adaptation mechanisms. Data for autonomic balance, espe-

Table 6.6. Hemodynamic indices of the participants after completion of the trek (Ottawa)

Participant	Heart rate bpm	Arterial pressure, mm Hg		Pulse pressure mm Hg	Mean driving pressure mm Hg	Stroke volume ml/beat	Cardiac output ml/min	Kerdot's autonomic index %
		systolic	diastolic					
D.I. Shparo	73	130	80	50	96	55	3,989	−11
F.F. Konyukhov	86	110	80	30	90	42	3,604	6
V.I. Shishkarev	69	115	70	45	85	62	4,271	−1
M.G. Malakhov	55	120	85	35	97	39	2,212	−52
A.V. Melnikov	60	160	100	60	120	41	2,472	−67
A.A. Belyaev	80	130	80	50	97	54	4,342	1
A.P. Fedyakov	68	160	100	60	120	41	2,802	−47
Yu.I. Khmelevsky	82	115	70	45	85	53	4,332	12
V.P. Ledenev	90	150	100	50	117	34	3,062	−12
M. Buxton	66	115	70	45	85	60	3,967	−6
L. Dexter	65	130	80	50	97	59	3,974	−23
R. Weber	72	110	70	40	83	61	4,468	14
Mean ± SD	72 ± 7	129 ± 14	82 ± 20	47 ± 15	98 ± 25	50 ± 11	3,616 ± 1,232	15 ± 41

Table 6.7. Cardiac rhythm indices of the participants after completion of the trek (Ottawa)

Participant	Cardiac cycle, ms		Coefficient of variation %	Mode, cardiac cycle ms	Amplitude about mode %	Range of variation ms	Stress index arbitrary units
	mean	SD					
D.I. Shparo	937	39	4.1	945	62.0	180	183
F.F. Konyukhov	855	27	3.1	825	56.0	140	244
M.G. Malakhov	976	34	3.5	995	61.0	200	153
A.A. Belyaev	935	31	3.3	905	56.0	120	257
Yu.I. Khmelevsky	905	24	2.7	935	55.0	100	292
R. Weber	900	61	6.8	905	54.0	520	57
K. Holloway	869	37	4.3	875	40.0	160	142
Mean ± SD	911 ± 42	36 ± 12	4.0 ± 1.4	912 ± 54	54.0 ± 7.6	203 ± 144	189 ± 81

cially the decreased variance of the heart rate, suggested that in five individuals, the regulatory mechanisms were more stressed than at the initial examination, with inadequate adaptation to the environment. In three of this group (D.I.S., Y.I.K., A.A.B.) changes were also more marked than at the 'fifth drop'. The best adapted, in terms of autonomic function, seemed to be R.W. and K.H.

Relationship to Mission Success

A final step in the analysis compared physiological data with individual success over the course of the mission; the basis of this assessment is discussed further in chapter 9.

Canonical analysis demonstrated strong relationships between expert appraisals of functional success and the parameters of cardiac rhythm. Association of success with the initial data is perhaps a legitimate exercise, although the authors note that there may have been some stress upon the subjects even during their first exmination. With respect to later data, it is even more possible that success may have influenced cardiac rhythm, rather than the converse.

Specifically, expert appraisals of success showed a significant positive correlation with respiratory constituents of the cardiac rhythm spectrum, and a negative correlation with the slow wave constituents. This suggested to the Soviet cardiologists that the less the initial centralization of control processes and the less the activation of subcortical nervous centers and synmpathetic mechanisms over the course of the trek, the better the participants coped with the mission.

The expert appraisal, again discussed further in chapter 9, showed a significant positive correlation with the mode of cardiac cycle duration, and a negative correlation with the amplitude and the stress index. Thus the higher the initial functional stress (and, in Western thought, the lower the initial state of training), the lower the expert appraisal at the end of the mission, and the worse the individuals were seen to have coped during the crossing.

7 Physical Working Capacity and Body Composition[1]

M.A. Booth, J.S. Thoden, F.D. Reardon, M. Jetté,
J. Quenneville, A. Rode, V.S. Koscheyev, V.K. Martens,
T.S. Bashir-Zade, A.A. Belyakov, G.V. Kypor, A.V. Visochanski[2]

Anthropometric Measurements

The anthropometric measurements included: height; body mass; five skinfolds (triceps, biceps, subscapular, iliac crest and medial calf) [Fitness and Amateur Sport, 1986]; five girths (chest, waist, gluteal, thigh and calf) [Jetté, 1983a, b], and four diameters (biacromial, bicristal, bicondylar femur, and elbow breadth) [Jetté, 1983a, b]. Body density was assessed by hydrostatic weighing. In Ottawa, the subject was submerged in an aluminum cage suspended in a water basin and connected to a precision load cell (model 5341, GSE Inc., Farmington Hills, Mich., USA). Residual volume was estimated from vital capacity [Wilmore, 1969], as determined by a 13.5-liter Warren Collins spirometer. The percentage of body fat was calculated from the equation of Siri [1956]. In Moscow, the same procedures were used, except that the subject was seated in a weighted chair, and the load cell readings were taken directly from a voltmeter.

From the measured variables, the following indices were calculated: body mass index (BMI); waist-to-hip ratio (WHR); predicted percentage of body fat [Durnin and Womersley, 1974], and chest minus waist difference [Jetté, 1983a].

Muscle Strength and Lung Function

Muscle strength was assessed by a TKK handgrip dynamometer, and a shoulder-arm push dynamometer [Jetté et al., 1984].

Pulmonary function (vital capacity, VC; forced vital capacity, FVC: 1-second forced expiratory volume, $FEV_{1.0}$, and peak expiratory flow rate, PEF) was measured with a Vitalograph spirometer (S. Model 20,400, Roxon Medi-tech, Montreal, Que., Canada).

[1] Freely adapted from the Russian translation by R.J.S.
[2] Suzanne Cormier, Josée Quenneville, Josée Barsalou, Sylvia Weihrer, Michael Booth, and Nancy Saumure all assisted in the Canadian laboratory studies discussed in this chapter.

Cardiovascular Function

When the entire team was assembled in the USSR, cycle ergometry was completed on 12 members of the expedition. A Tunturi/Weider ergometer was used, two 5-min work bouts being separated by a rest interval of 3 min. The first load was chosen to induce a heart rate of 100–120 bpm by the end of the 5th min, while the second load was adjusted to yield a final heart rate of 150–170 bpm. The physical working capacity (PWC) was calculated according to the formula:

$$PWC_{170} = W_1 + (W_2 - W_1)(170 - f_{h1})/(f_{h2} - f_{h1}),$$

where W_1 and W_2 were the two work rates, and f_{h1} and f_{h2} were the corresponding heart rates. The corresponding maximal oxygen intake was derived from a reference table [Miroshnikov, 1971]. The systemic blood pressure was checked in the final minute at each loading and after 3 min of recovery.

The Canadian laboratory tested the PWC_{170} as in Moscow, and subjects were then told to continue cycling to exhaustion as the loading was progressively increased. Both before and after the trek, members of the Canadian team also completed a standard protocol for the determination of maximal oxygen intake. In this test, the subjects cycled at 60 rpm, with an initial loading of 2 Pa; this was increased in stages of 1 Pa, so as to exhaust the subjects in 6–7 loads. The oxygen consumption, expiratory volume and breathing frequency were measured using a Roxon Medi-tech System (Thermox Instruments Division, Pittsburgh, Pa., USA).

Comparisons were made between the Canadian and Soviet assessments of power output, and it appeared that there were no significant problems from inter-laboratory differences in the calibration of cycle ergometers [Kane and Jones, 1982].

Three members of the Canadian team were tested further, using a motorized ski-treadmill. They were initially allowed to familiarize themselves with the ergometer and with the mouthpiece, and then definitive measurements were made with and without a 37.5-kg backpack, both shortly before and 1 week after completion of the trek. Standard metabolic techniques were used to assess the oxygen cost of skiing, the subjects being tested for 5 min at each of six speeds from 2.5 to 5.0 km/h at a constant 1% slope, these speeds being chosen to match the anticipated pace of skiing on the expedition. The heart rate was monitored by battery-operated Sports Testers (Noram Patient-Care Products). Expired gas was chanelled through a Roxon metabolic cart system, with Applied Electrochemistry oxygen and carbon dioxide analyzers and a Morgan ventilation monitor. The final 2 min at each speed was used to estimate the oxygen cost of that stage.

During the trek, heart rates were measured by three Sports-Tester heart rate monitors, shared among Canadian and Soviet team members. The chest straps and watches were donned in the morning, and were worn all day by the selected subjects, values being sampled and stored every 15 min throughout the day. Useful data were obtained from five participants; the number of days logged ranged from 9 to 28 for a total of 84 man-days of data. The findings were related to graphs of heart rate against oxygen consumption obtained in the base laboratory. The logged heart rates were used to represent the daily activity profile, and the range from minimum to maximum work-rate.

Anthropometric Data and Muscle Strength

Data were obtained on all 13 participants. Because of differences in mean age, results were analyzed separately for the Soviet and the Canadian participants.

The Soviet team, with an average of 40.9 ± 7.2 years, had a height of 177.1 ± 5.8 cm, an initial body mass of 80.6 ± 8.4 kg, and a BMI of 25.7 ± 2.0 units (table 7.1). The Canadians had a mean age of 33 ± 6.8 years, with a height of 178.1 ± 8.2 cm, a body mass of 74.8 ± 5.6 kg, and a BMI of 23.6 ± 1.1 units. The Soviets also carried more body fat than the Canadians (24.5 versus 8.6%).

During the first month of the trek, the body mass of the Soviet participants decreased by 4–5 kg (fig 7.1). It then increased by 3–4 kg, so that by the end of the expedition their mean body mass was 2.2 kg less (2.7%) than at the outset. For the Canadians, the mean body mass dropped during the first 2 weeks of the expedition (2 kg, 2.7%). It then increased as the expedition progressed, so that their final body mass averaged 75.6 kg, an increase of 0.8 kg (1.6%) over the outset, with little increase in the percentage of body fat (1.2%), implying that there had been an increase of up to 1.2 kg in lean tissue mass. Figure 7.1 also illustrates the weight change that occurred during the rest session of May 13–14. The mean body mass of the group increased by 2 kg during this short period.

The changes in triceps skinfold over the trek are illustrated in figure 7.2. Although the mass of the skiers dropped over the first 2 weeks, the skinfold readings actually increased slightly during the trek, reaching their highest level midway through the trek, with a small decrease thereafter.

The mean combined grip strength decreased slightly in the Soviet participants, from 1,138 N before the trek to 1,118 N afterwards. In the Canadians, the average value remained stable at 1,158 N (table 7.2). The

7 Physical Working Capacity and Body Composition

Table 7.1. Age and anthropometric characteristics of participants

Variables	Canadians		Soviets		Group	
	before	after	before	after	before	after
Age, years	33.0	–	40.9	–	38.5	–
Height, cm	178.1	177.6	177.1	177.2	177.4	177.3
Body mass, kg	74.8	75.6	80.6	78.4	78.8	77.6
Chest girth, cm	94.4[a]	93.1[a]	101.4[a]	98.0[a]	99.2*	96.5*
Waist girth, cm	80.0[a]	79.9[a]	87.9*[a]	84.0[a]*	85.5*	82.8*
Gluteal girth, cm	95.5	94.2	99.1*	96.7*	98.0*	95.5*
Thigh girth, cm	56.9	–	56.8	–	56.8	–
BMI, kg/m^2	23.6	24.0	25.7	25.0	25.1	24.7
Body fat, %	8.6[a]	8.7[a]	24.4*[a]	19.4*[a]	19.6*	16.1*
Chest to waist, cm	14.4	13.3	13.4	13.9	13.7	13.7
Waist/Hip ratio	0.84	0.85	0.89	0.87	0.87	0.86

* Before – after significant difference at $p < 0.05$; [a] = Significant difference between Canadians and Soviets at $p < 0.05$, before or after trip

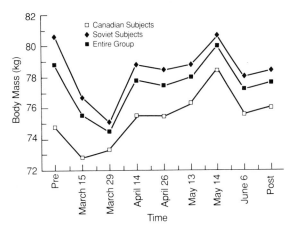

Fig. 7.1. Body mass over course of the trek.

upper body strength increased from 776 to 800 N in the Soviet participants, and from 765 to 800 N in the Canadians, these changes being compatible with the increments in lean mass.

The early loss of body mass was probably due to dehydration, a result of a restricted water intake and diarrhea. The intake of fat was high at the beginning of the trek, causing gastrointestinal problems (particularly for

Table 7.2. Fitness characteristics of the subjects

Variables	Canadians		Soviets		Group	
	before	after	before	after	before	after
Combined grip force, N	1,158	1,158	1,138	1,118	1,148	1,128
Upper arm force, N	765*	800*	776	800	773	800
PWC, W	302[a]	250	258*[a]	218*	318*	228*
VO_{2max}, ml/kg/min	60.5	48.7	–	46.2	–	47.1
Maximum heart rate, bpm	188	184	–	186	–	185

* Before – after significant difference at $p < 0.05$. [a] Significant difference between Canadians and Soviets at $p < 0.05$.

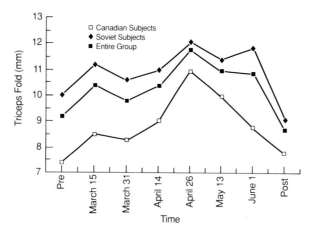

Fig. 7.2. Triceps skinfold thickness over the trek.

the Canadians, who were not accustomed to this type of diet). Later changes reflect rehydration and muscle hypertrophy. On average, there was a 21% decrease in body fat (from 24.5 to 19.4%), suggesting a 4.5-kg reduction in body fat and a 2.3-kg increase in lean tissue in the Soviets.

Cardiovascular Data Obtained on Canadian Team

All four Canadians completed determinations of aerobic power before and after the trek. However, subject D could not schedule the initial

7 Physical Working Capacity and Body Composition

Table 7.3. Cardiovascular responses to cycling

Variables	Canadians		Soviets		Group	
	before	after	before	after	before	after
Resting						
Heart rate, (bpm)	–	84	–	82	–	82
Blood pressure, mm Hg						
Systolic	117	111	128	125	125	121
Diastolic	73	71	83	77	80	75
Load A						
Heart rate, bpm	112	128	108*	132*	109*	131*
Blood pressure, mm Hg						
Systolic	142	159	164	183	159*	175*
Diastolic	73	64[a]	79	78[a]	78	74
Load B						
Heart rate, bpm	168	169[a]	170	178[a]	170	175
Blood pressure, mm Hg						
Systolic	187	181	201	209	198	201
Diastolic	70*	63*[a]	76	84[a]	74	78
Maximum Effort						
Heart rate, bpm	–	184	–	186	–	185
Blood pressure, mm Hg						
Systolic	–	185	–	209	–	200
Diastolic	–	68[a]	–	79[a]	–	75

* Before – after significant difference at $p < 0.05$.
[a] Significant difference between Canadians and Soviets at $p < 0.05$.

ski-treadmill tests, and in subject C the initial tests without backpack were considered invalid, since he attempted to 'toe-kick and glide' rather than 'trudge', resulting in energy cost values that were almost as great as with a backpack.

Maximal aerobic power was significantly reduced in all four team members (average decrease 20%). The initial maximal oxygen intake averaged 60.5 ± 5.1 ml/kg · min, dropping to an average of 48.7 ± 4.9 ml/kg · min at the final evaluation. The Soviet team also participated in this final evaluation, with values of 46.2 ± 5.5 ml/kg · min.

The resting and exercise heart rate and blood pressure measurements before and after the trek are shown in table 7.3. The mean systolic blood pressure of the Soviets before the trek (128 ± 17 mm Hg) was higher than that of the Canadians (117 ± 8 mm Hg). This was also true of diastolic pressure (83 vs. 73 mm Hg). These resting pressures decreased slightly

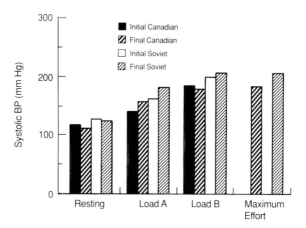

Fig. 7.3. Systolic blood pressure during cycling test.

following the trek. However, the mean group exercise systolic pressure at the first exercise load increased significantly from 159 to 175 mm Hg from before to after the trek. At the second exercise load, the Canadians had slightly lower pressures than before the trek, while the Soviet participants again showed slightly higher pressures, achieving a maximum systolic pressure of 209 mm Hg at this load (fig. 7.3). The mean maxima at the second load for Canadian and Soviet teams were 209/84 versus 181/63 mm Hg. The increases in blood pressure could possibly be attributed to a combination of a restricted fluid intake and a high-salt diet. The Soviet scientists also observed an increase in circulating aldosterone levels at the beginning of the trek; aldosterone has a direct effect upon the kidneys, increasing the retention of sodium ions and water, with an increase in extracellular fluid volume.

The average (monthly) heart rate recorded by the Sports Testers decreased gradually in all five subjects during the course of the trek (table 7.4). For example, in M.B., the average heart rate was 129 bpm at the beginning of the expedition, but 119 bpm at the end. This skier reached a heart rate of 150 bpm on day 2 when skiing for 3.8 h under very cold and hazardous conditions, but an average of 106 bpm was seen when covering flat ice in late May. The oldest skier (L.D., 43 years) showed the smallest decrease in heart rate over the trek (2%), while the youngest participant (R.W., 28 years) had the largest decrease (8%).

The decrease in fitness of the Canadians might be expected from a 3-month period when the intensity of exercise was substantially less than in training before the trek (5–6 days/week for several months at heart rates >140–150 bpm). The heart rate data from subject B (fig. 7.4) is fairly

7 Physical Working Capacity and Body Composition

Table 7.4. Average heart rates (bpm) during the ski trek

Skier	March	April	May
M.B.	128.5	125.4	119.2
R.W.	120.8	117.9	111.6
M.M.	128.3	123.5	121.0
L.D.	126.0	127.5	123.6
C.H.	–	114.8	108.9
Mean	125.9	121.8	116.9
SD	3.6	5.3	6.3

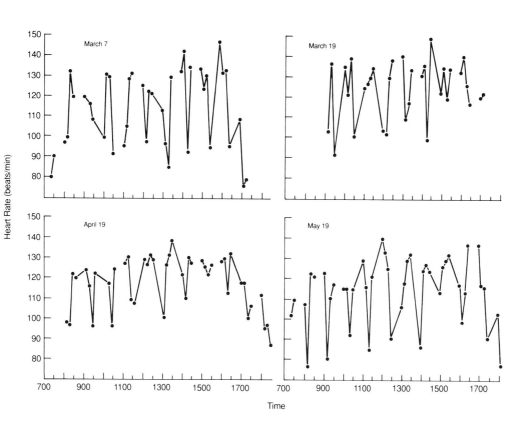

Fig. 7.4. Polar trek heart rates for subject B during trekking. Data on 4 representative days of the 3-month expedition.

typical of the 'stress' experienced at four different times during the traverse, and is consistent with the hypothesis of 'detraining' during the trek, with activity heart rates falling most consistently in the 120–130 bpm range. This view is further substantiated by hormonal data for the Soviet team members. Levels of cortisol during preparations for the trek were two times higher, and insulin was 30% lower than the normal values established over a number of years, but during the trek these values returned toward normal values.

However, it is also possible that the observed decrease in fitness scores could be attributed in part to the lack of specificity of the test format (cycle ergometer for skiers), with added contributions from generalized fatigue after the trek (although maximal heart rates were very similar before and after the expedition).

The ski treadmill tests were submaximal, and were completed without difficulty by all subjects; the increases in speed did not induce any noticeable change in skiing technique, which could best be described as 'ski-walking' or 'trudging'. At the initial testing, heart rate and oxygen consumption increased progressively with an increase in treadmill speed (fig. 7.5), but the oxygen cost per kilometer travelled remained relatively constant over the range of speeds tested. The backpack, which increased the total mass by an average of about 50% (table 7.5), increased the oxygen cost on the ski-treadmill by 30% in two subjects, and by 49% in the third. Following the expedition, these measurements were repeated. Given the significant detraining, a logical expectation would have been for a higher heart rate after the trek at any given submaximal exercise intensity. However, this was not the case (fig. 7.5). At any given velocity, the heart rates were unchanged in subjects A and B, and were actually 8% lower in the loaded test for subject C (table 7.6). The explanation is seen in the oxygen consumption data. In all three subjects with complete data sets, the oxygen cost of loaded skiing had decreased at all the speeds tested, by an average of 6, 8 and 14% respectively; subjects A and B showed similar decreases in the cost of unloaded skiing. These results suggest that there was a significant improvement in the economy of movement, sufficient to compensate entirely for the coincident decreases in fitness. Moreover, the decreases in oxygen cost were apparently greatest at the speed of 3.5 km/h (14, 7 and 23% for the three loaded subjects, table 7.5), suggesting some degree of specificity in the adaptation.

Another point of interest is that within any one individual, the oxygen cost per kilometer traversed was relatively constant over the mid-range of speeds (fig. 7.5). This is consistent with other forms of low-speed human locomotion in which air resistance is not a significant factor [DiPrampero, 1986]. However, there was a rather wide interindividual variation in both

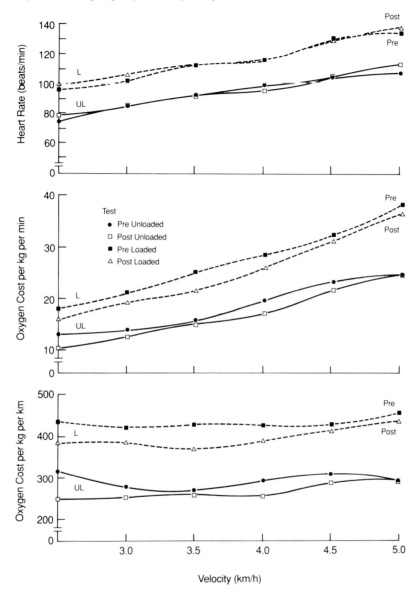

Fig. 7.5. Heart rate and oxygen consumption responses of subject A to loaded (L, 37.5-kg pack) and unloaded (UL) treadmill ski trekking before and after the 3-month transpolar expedition.

Table 7.5. Body mass of Canadian participants

Subject	Body mass, kg		Pack load[1] as % of body mass before trek
	before	after	
A	68.8	70.6	54.5
B	73.0	75.6	51.4
C	82.3	85.7	45.6
D	75.0	72.1	50.0
Mean	74.8	76.0	50.4

[1] Pack load of 37.5 kg.

Table 7.6. Economy and relative stress of trekking in Canadian participants

Subject	VO_2, ml/kg per km			Heart rate, bpm		% $\dot{V}O_{2max}$	
	before	after	after : before	before	after	before	after
A	429	370	0.86	112	112	37	39
B	447	417	0.93	126	126	42	46
C	422	324	0.77	114	99	43	41
D	–	389	–	–	129	–	46
Mean	433	370	0.86	117	112	40.7	42.0

Subject 4 was excluded from the means. Values are for the 3.5-km/h loaded condition on the ski treadmill.

economy and relative stress of trekking at any given speed (table 7.6). In the final loaded tests at 3.5 km/h, the oxygen cost (ml/kg per km) was still 29% greater in the least than in the most efficient member of the expedition. Even if oxygen cost is expressed in terms of the total mass carried (body plus pack), this difference is only reduced to 24%. This factor may deserve consideration, along with the oxygen consumption at the anaerobic threshold, when choosing a participant to serve as pacesetter on such expeditions.

The objective before departure was to ski for about 10 h/day, resting for the last 10 min of each hour, with 3-day layovers every 10 days. This would have allowed completion of the 1,730-km route in about 600 h of actual travel at a speed of 2.9 km/h. However, on some days ice and wind conditions limited the distance to 3–5 km, with only an hour of travel, while on more favorable days the journey time was extended to 12 h. When

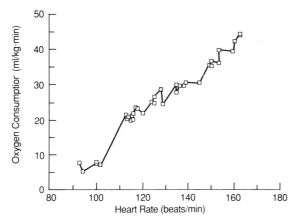

Fig. 7.6. Laboratory calibration (oxygen consumption for given heart rate when using ski treadmill prior to expedition).

these limitations imposed by weather, surface, terrain and detours are considered, the actual speed seems consistent with the ski-treadmill estimate of 3.5 km/h for stretches of flat and even terrain, although it is recognized that the laboratory calibration (fig. 7.6) does not give a precise measure of the energy cost of wearing full arctic gear while climbing over 6-meter pressure ridges at a temperature of $-40\,°C$ on a windy day. In the laboratory, this speed demanded an average oxygen consumption of 21.6 ml/kg · min (370 ml/kg per km of distance covered, 32–39% of maximal oxygen intake) at the conclusion of the trek. When the estimated basal costs are subtracted and the energy cost is calculated at a rate of 20.2 kJ/l of oxygen (RQ = 0.82), the cost of the loaded trekking itself (for an average of 8.3 h/day) calculates to an additional energy need of about 11.3 MJ/day, essentially twice the energy needed on non-trekking days. However, it should be stressed that this estimate is based on the heart rate/oxygen consumption line measured in the laboratory, and that needs could be further augmented by cold exposure.

Ergometric Data from Moscow

Initial Results

Initial measurements were made in Moscow, shortly before departure on the ski-trek. Taking account of the individual ages, which ranged from 28 to 51 years, all subjects were in good physical condition, the highest scores being attained by M.B., R.W., D.S. and V.I.S., and the lowest

Table 7.7. Physical working capacity (PWC_{170}) of the participants before the expedition (Moscow)

Participant	Age years	Work load I W	Work load II W	f_{h1} bpm	f_{h2} bpm	PWC_{170} W	$PWC_{170'}$ W/kg
D.I. Shparo	46	25	54	98	171	309	3.5
F.F. Konyukhov	36	19	43	119	180	235	3.3
V.I. Shishkarev	39	17	37.5	94	150	307	3.8
M.G. Malakhov	34	17.5	42	103	171	249	3.3
A.V. Melnikov	45	17	37.5	97	160	245	2.9
A.A. Belyaev	29	21	46	116	180	251	3.4
A.P. Fedyakov	48	17.5	40	112	177	226	2.8
Yu. I. Khmelevsky	51	23	44	114	170	315	3.3
V.P. Ledenev	41	17	36	120	168	220	2.8
M. Buxton	31	21	46	106	168	270	3.7
L. Dexter	43	20	50	113	174	288	3.9
R. Weber	28	25	50	118	162	327	4.8
Mean ± SD						265 ± 32	3.4 ± 0.6

results by V.P.L., A.P.F. and A.V.M. As expected, the younger participants tended to achieve higher results, both in absolute and relative terms (table 7.7). The corresponding estimates of maximal oxygen intake (liter/min) were: R.W. 5.25; L.D. 4.37; V.I.S. 5.15; F.F.K. 3.92; M.G.M. 4.34; Y.I.K. 5.17; V.P.L. 4.59; A.V.M. 4.01 and A.P.F. 4.12.

The systemic blood pressures observed during the test were generally appropriate to the applied ergometer loading (table 7.8). The largest reactions at the second loading were shown by V.I.S., A.P.F. and Y.I.K. Two of these subjects also had high resting pressures, their reactions being pathologic rather than physiologic. The lowest exercise pressures (M.G.M., R.W. and V.P.L.) were within the normal range.

Final Results

In agreement with the Canadian laboratory data, the group average PCW_{170} showed a significant ($p < 0.05$) decline by the end of the expedition (table 7.9), from 265 to 242 W (3.4 dropping to 3.0 W/kg). However, some subjects showed no change (F.F.K, V.I.S., M.G.M., Y.I.K., V.P.L., R.W.), in one of the least fit (A.P.F.) there was an increase, and in four individuals, three of whom were fitter members of the party, there was a decrease (L.D., M.B., D.I.S. and A.V.M.).

Blood pressure responses after the expedition (table 7.10) were generally unchanged relative to the initial responses.

7 Physical Working Capacity and Body Composition

Table 7.8. Arterial pressures (mm Hg) of the participants during the PWC test before the expedition (Moscow)

Participant	Resting		5th min of load I		5th min of load II		3rd min of recovery	
	systolic	diastolic	systolic	diastolic	systolic	diastolic	systolic	diastolic
D.I. Shparo	135	70	185	80	210	75	150	70
F.F. Konyukhov	115	80	170	80	185	85	190	90
V.I. Shiskarev	140	90	190	85	240	90	180	75
M.G. Malakhov	100	70	140	65	170	70	135	65
A.V. Melnikov	130	60	185	75	210	75	150	70
A.A. Belyaev	125	70	195	75	220	70	135	60
A.P. Fedyakov	155	100	215	90	230	90	175	85
Yu.I. Khmelevsky	115	70	200	80	235	105	235	105
V.P. Ledenev	120	80	165	75	185	90	170	80
M. Buxton	115	70	175	65	190	60	145	70
L. Dexter	125	80	150	85	190	70	130	75
R. Weber	110	80	155	50	195	65	160	55
Mean ± SD	124 ± 15	77 ± 11	177 ± 22	76 ± 9	205 ± 23	80 ± 13	163 ± 30	75 ± 14

Table 7.9. Physical working capacity (PWC_{170}) of the participants after completing the trek (Ottawa)

Participant	Age years	Work load I W	Work load II W	f_{h1} bpm	f_{h2} bpm	PWC_{170} W	$PWC_{170'}$ W/kg
D.I. Shparo	46	25	54	153	179	209	3.0
F.F. Konyukhov	36	19	43	130	177	204	2.8
B.I. Shishkarev	39	17	37.5	150	171	275	3.6
M.G. Malakhov	34	17.5	42	145	174	208	2.5
A.V. Melnikov	45	17	37.5	162	186	176	2.3
A.A. Belyaev	29	21	36	120	179	228	2.85
A.P. Fedyakov	48	17.5	40	112	177	309	2.8
Yu.I. Khmelevsky	51	23	44	108	174	285	3.3
V.P. Ledenev	41	17	36	120	182	240	3.1
M. Buxton	31	21	46	109	171	249	3.2
L. Dexter	43	20	50	131	185	236	3.15
R. Weber	28	25	50	124	176	252	4.0
Mean ± SD						242 ± 40	3.4 ± 0.5

Table 7.10. Arterial pressures (mm Hg) of the participants during the PWC test after completing the expedition (Ottawa)

Participant	Resting		5th min of load I		5th min of load II		3rd min of recovery	
	systolic	diastolic	systolic	diastolic	systolic	diastolic	systolic	diastolic
D.I. Shparo	115	70	175	80	210	80	185	70
F.F. Konyukhov	120	80	145	70	185	70	150	75
V.I. Shishkarev	130	80	135	70	185	70	140	80
M.G. Malakhov	115	70	145	90	170	70	145	80
A.V. Melnikov	120	85	140	75	190	70	140	70
A.A. Belyaev	130	80	160	75	180	75	140	85
A.P. Fedyakov	160	100	220	90	240	95	230	90
Yu.I. Khmelevsky	150	100	215	90	240	90	170	90
V.P. Ledenev	110	80	145	70	215	60	150	80
M. Buxton	115	70	140	70	200	70	130	70
L. Dexter	125	80	150	85	190	70	130	75
R. Weber	110	70	136	65	170	65	130	70
Mean ± SD	126 ± 15	80 ± 12	160 ± 26	85 ± 10	198 ± 32	74 ± 14	155 ± 16	77 ± 12

Correlation with Expert Appraisals

Expert appraisal of mission success, as discussed in chapter 9, showed a substantial positive correlation with the PWC_{170} per kilogram of body mass. The greater the individual's working capacity, the higher the expert appraisal and the greater the likelihood of successful participation in the mission. The three most efficient individuals were judged to be R.W., V.I.S. and M.G.M., while the least efficient were A.V.M., A.P.F. and L.D. (the last, paradoxically, had a high physical working capacity, but had only limited experience of skiing).

But despite the large and culturally diverse nature of the team, the harsh climatic conditions and the isolation of the north, every one of the 13 skiers succeeded in completing the trek. The average age of the group was over 40 years, and five skiers were older than 43 years, the oldest (Y.K.) being 51 years. The group included individuals with excessive body mass, hypertension, and chronic obstructive lung disease. One man was missing the front third of one foot due to freezing in an earlier expedition. With a few exceptions, the participants were not of high aerobic fitness. However, they did possess a high degree of muscular endurance and stamina. Thus it would seem worthwhile to reevaluate the accepted criteria for screening and selecting candidates who must work extremely hard under extreme environmental and psychological conditions.

8 Chamber Simulation of Ski Trek[1]

B.A. Utehin, M.G. Malakhov, M. Jetté, A. Rode,
S. Livingstone, R.W. Nolan, A.A. Keefe

Introduction

The conditions likely to be encoutered during a typical day of the ski trek were simulated in two micro-climatic chambers (TBVK-5-17 and TBVK-5-8) at the USSR Ministry of Public Health Institute of Biophysics. Parallel investigations were completed in Ottawa, at the experimental facilities of the Defence Research Laboratories, after completion of the expedition. Observations on local acclimatization of the hands were also undertaken on Canadian participants.

Biochemical reactions to the chamber simulations are discussed in chapter 11.

Methodology

Soviet Studies
After an appropriate control examination, subjects spent a period of 5-6 h overnight in the climatic chamber at a temperature of $-20\,°C$ and a relative humidity of 40-60%. Observations were made before and after sleeping. The subjects then undertook three bouts of physical activity on a step test, climbing at a pace of 15 steps/min. Each bout of activity lasted 50 min, with a 10-min recovery period between bouts of exercise.

As during the expedition, subjects were equipped with sleeping bags, anoraks, cotton and woollen underwear and other appropriate clothing. A 40-kg rucksack was carried during the stepping exercise.

Skin temperatures were measured at five sites, and rectal temperatures were also recorded, allowing computation of an average weighted skin temperature (according to the formula of N.K. Witte) and an average body temperature.

Cardiorespiratory responses to the chamber simulations were evaluated by electrocardiography, pulsometry, and measurements of respiratory rate, respiratory minute volume and oxygen consumption.

[1] Freely adapted from the Russian translation by R.J.S.

Psychological status was evaluated by self-appraisal indicators (the SAM test) and by measurement of simple and complex visuomotor reaction times.

Canadian Studies

The experiments to assess general adaptation to cold were initiated in a laboratory at a temperature of $23 \pm 1\,°C$. Each subject, wearing only shorts, had a rectal thermistor inserted 15 cm into the rectum, and YSI type 44004 thermistors attached to the skin using Blenderm surgical tape (3M Company) at nine sites (forehead, forearm, back of hand, top of foot, calf, thigh, abdomen, middle of left finger, and big left toe). Three disposible monitoring electrodes (Hewlett Packard HP 14445A) were attached to the skin in the areas of the upper arm muscles and the thigh muscles. The subject then entered a single-layer Canadian Forces sleeping bag and reclined supine on a rope-mesh cot fitted with an oronasal face mask (Speakeasy). The cot and subject were moved from the laboratory into an environmental chamber maintained at $10 \pm 0.5\,°C$.

In the environmental chamber, the facemask was attached to a Beckman metabolic cart, and leads from the skin electrodes were attached to a Beckman R511A Dynagraph with electromyograph couplers (Beckman 9852A). The sleeping bag was removed after a control period of 30 min, and the subject remained in the environmental chamber for an additional 60 min.

Throughout the experiment, the subject's metabolic rate was measured and recorded every minute, electrical activity from the arm and thigh muscles was monitored continuously, and all temperatures were measured and recorded every minute using an automated data acquisition system (HP 3497A Data Acquisition/Control Unit and HP 85 computer). Mean skin temperatures (Ts) and mean body temperatures (Tb) were also calculated every minute, using the equations of Hardy and Dubois [1938] and Burton [1935], respectively:

$$Ts = 0.07\,(\text{forehead}) + 0.14\,(\text{forearm}) + 0.05\,(\text{back of hand})$$
$$+ 0.07\,(\text{top of foot}) + 0.13\,(\text{calf}) + 0.19\,(\text{thigh}) + 0.35\,(\text{abdomen})$$

$$Tb = 0.67\,(\text{rectal}) + 0.33\,(Ts).$$

Local cold adaptation was determined by measuring cold-induced vasodilatation (CIVD) in the finger, caused by immersing it in a stirred ice-water bath (temperature between 0 and $0.1\,°C$). A thermistor (YSI 44004) was attached to the pad of the mid-finger of the nondominant hand of each subject, using Blenderm surgical tape. A thin latex surgical glove was then donned to prevent water from penetrating between the thermistor and the skin. After several minutes at ambient temperature ($23 \pm 1\,°C$), the finger was immersed in ice water for 30 min. Throughout the experiment, the temperature of the finger was measured every 5 s, using the automated data acquisition system.

Measures of local and general cold adaptation were completed twice, 1 month before the trek and within 5 days of return from the Canadian Arctic.

Results

Soviet Studies

Thermal data (table 8.1) showed initial rectal temperatures ranging from 36.4 to 37.5 °C. During sleep, temperatures decreased by 0.2–1.0 °C, but the 50-min bout of physical exercise was sufficient to induce a sharp increase of core temperature (0.7–1.4 °C), associated with an average exercise heart rate of 132 bpm. Stabilization of the body temperature occurred during the second and third bouts of exercise, further increments in rectal temperature amounting to only 0.1–0.3 °C. With the exception of V.I.S. (rectal temperatures of 36.4, 36.6 and 37.2 °C in the three work bouts), the rectal temperatures while working did not drop below 36.9 °C.

During sleep, the weighted average skin temperature increased sharply, reaching values of 32.5–35.0 °C in different subjects, suggesting that the protective garments provided adequate insulation while sleeping. The average body temperature of most subjects increased by 0.2–0.6 °C while sleeping, although in two subjects it remained stable, and in one subject there was an 0.6 °C decrease. Nevertheless, night-time temperatures remained essentially in the optimal range.

During exercise, there was a decrease in the average weighted skin temperature despite the rise in rectal temperatures. By the end of the first hour of work, the skin temperatures in six of the participants (A.P.F., D.I.S., V.I.S., M.B., Y.I.K. and A.I.M.) were around threshold values, and by the end of the 3rd h of physical activity readings of 28–29 °C were imposing a significant stress on thermoregulatory mechanisms. In the remaining subjects, skin temperatures also decreased while working, but nevertheless remained within the permissible range. In two subjects (D.I.S. and Y.I.K.), the average body temperature dropped to undesirably low levels (33.6 °C at the end of the 3rd h of work, and 34.2 °C at the end of the 2nd h of work, respectively).

Table 8.1. Thermal data (°C, mean ± SD) for 12 participants during the simulation experiment before the expedition (Moscow)

Stage	Skin temperature		
	forehead	chest	wrist
Before sleep	33.5 + 0.9	31.8 + 2.0	31.8 + 1.1
After sleep	32.5 + 2.4	34.8 + 1.0	33.4 + 2.5
After the 1st h of physical work	29.9 + 3.4*	32.1 + 2.3*	25.4 + 3.1*
After the 2nd h of physical work	29.2 + 2.8*	31.0 + 3.4*	24.7 + 4.1*
After the 3rd h of physical work	29.1 + 2.7*	30.8 + 4.4*	22.3 + 5.3*

* Significance of difference relative to reading 'after sleep': $p < 0.05$.

Adjustments to the cold environment were primarily of the insulative type, heat loss being reduced by a marked vasoconstriction and a sharp drop in 'shell' temperature. Cooling of the extremities was particularly noticeable. For example, the wrist temperature by the end of the 3rd h of exercise was in the range of 16.8–24.7 °C.

Comparisons of the Soviet and Canadian clothing assemblies showed that both provided adequate insulation. However, the Canadian garments proved somewhat more effective than the Soviet counterparts, particularly when the subjects were exercising.

We may conclude that in general, the protective equipment provided was appropriate to the environment and the contemplated work rate, and that the temperature of the participants remained within acceptable limits during chamber simulations of the trek.

Canadian Studies

The results are shown in table 8.2 and figures 8.1–8.4. Because of scheduling and other prior commitments, not all the tests could be completed on all of the team members following the trek. However, the metabolic rate during the test of general cold tolerance was lower after than before the trek (fig. 8.1). This finding agrees with the electromyograph data which showed a delayed onset of shivering in three of the four Canadians (table 8.2). The remaining subject showed negligible shivering during exposure before the trek and no shivering during the experiment following the trek, which was terminated at 30 min.

There were no differences in skin, rectal or mean body temperatures between the initial and final tests (fig. 8.2), two of the subjects showing lower and two higher temperatures during the second exposure. The

		Rectal temperature	Weighted average skin temperature	Average body temperature
thigh	shin			
31.9 + 1.6	32.0 + 1.7	36.9 + 0.4	32.0 + 1.6	36.3 + 0.4
33.8 + 1.0	33.0 + 2.0	36.2 + 0.4	34.1 + 1.0	35.5 + 0.5
29.4 + 5.9	37.8 + 3.1	37.3 + 0.4	31.0 + 2.2	35.4 + 0.8
28.8 + 4.8	30.4 + 2.6	37.3 + 0.5	31.0 + 2.6	35.2 + 1.0
28.9 + 4.7	31.3 + 3.5	37.5 + 0.4	30.9 + 3.7	35.4 + 1.1

Table 8.2. Time to onset of shivering

Subject No.	Onset of shivering, min			
	Arm		Leg	
	before	after	before	after
1	42	52	1	14
2	48	–	30	–
3	16	19	1	18
4	7	29	5	29

temperature of the extremities (toe and finger) did not indicate any definite change in the rate of cooling (fig. 8.3).

The measurable factors suggested by Yoshimura and Iida [1950], namely the time of the first temperature rise, the temperature at the first temperature rise, and the average temperature from 5 to 30 min) all show a greater CIVD response after completion of the ski trek (fig. 8.4).

Changes in Thermal Reactions

The Canadian data show that after 90 days of exposure to cold, the skiers showed a hypothermic, insulative type of adaptation, with a smaller increase in metabolism and a delayed shivering response. The results correspond with the observations of Budd [1962] after a stay in Antarctica, Davis [1961] in men artificially acclimatized in a cold chamber, and LeBlanc [1956] who examined soldiers living in the Arctic; however, they differ from those of Rivolier et al. [1988], probably because the cold

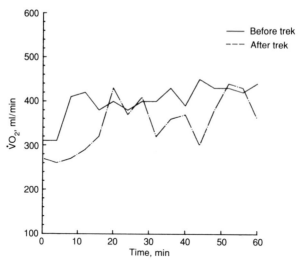

Fig. 8.1. Changes in the metabolic rate of one subject during general cold exposure.

exposure was much greater. The responses also differ from the increase in metabolic rate observed in various indigenous groups such as the Inuit of North America and the Indians of Tierra del Fuego [Hammel, 1964] and Norwegian youths who lived under primitive conditions on a mountain plateau of south-central Norway for a 6-week period in the autumn [Scholander et al., 1958].

The observation of an increased CIVD response agrees with many previous studies such as observations on the Inuit of North America [Meehan, 1954] and others exposed to cold for only a limited time [Scholander et al., 1958]. Yoshimura [1960] and Eagan [1963] found that resistance to cooling was also increased after repeated immersion of the toes or fingers in ice water over periods of several weeks. In the latter experiments, only the extremities and not the whole bodies of the subjects were cold-stressed. Other studies of persons who have worked in the cold for many years (for example, workers in cold chambers at −20 to −25 °C [Tanaka, 1971], Gaspé fishermen [LeBlanc et al., 1960] and British fish filleters [Nelms and Soper, 1962] again show an increased CIVD relative to control groups.

In contrast, an increased vasoconstrictor tone was found by Hampton [1969] in the hands of people exposed to cold in Antarctica, and Wyndham et al. [1964] noted that men who had lived in Antarctica for 1 year had lower hand temperatures when exposed to air temperatures of 5 °C. Decreased CIVD was also observed in soldiers after a 2-week Arctic military

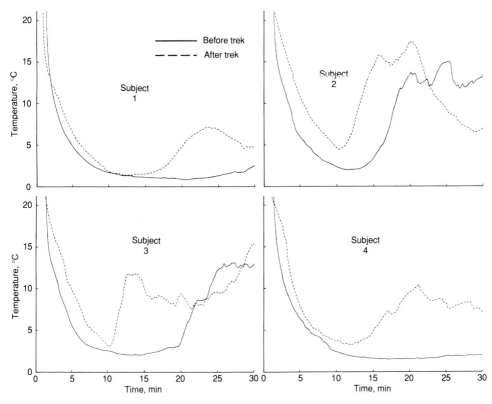

Fig. 8.2 Changes in the mean body temperatures of four subjects. Vertical lines denote standard deviation. Upper lines indicate rectal temperature changes, the middle lines mean body temperature and the lower lines mean skin temperature.

exercise. It was suggested that in this group either a deficiency of vitamin C [Livingstone, 1976a] or insufficient exposure to cold accounted for the decreased CIVD reaction. Recent studies of the cold adaptation of a group after participating in the International Biomedical Expedition to the Antarctic found no evidence of whole body acclimatization, but did find an enhanced vasoconstrictor response in the hands on exposure to ice water [Rivolier et al., 1988]; in this experiment, the effects of any acclimatization may have decreased during the rather long time that elapsed between the end of the expedition and the testing of the subjects.

The increased CIVD observed in our experiments suggests that after the trek the extremities would cool less during general exposure to cold. On the other hand, the whole-body cooling experiments indicate little or no difference in cooling of the extremities after the trek. Keatinge [1957] has

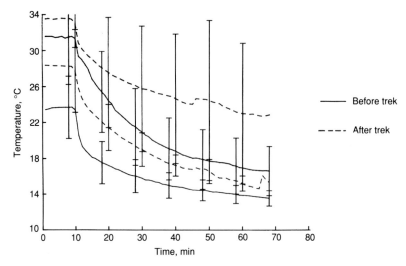

Fig. 8.3. Mean changes in finger and toe temperatures during the general cold exposure. Vertical lines indicate standard deviation. The upper two lines denote finger and the lower two lines denote toe temperatures.

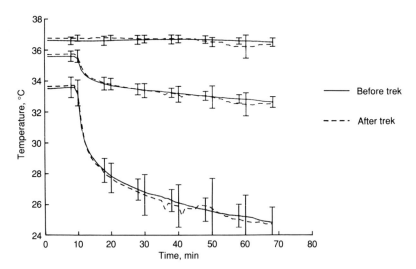

Fig. 8.4. Changes in finger temperature of each subject during finger immersion in ice water.

Table 8.3. Heart rate and respiratory variables during climatic chamber simulation of the ski trek in Moscow (mean ± SD)

Stage	Heart rate bpm	Respiratory rate breath/min	Respiratory minute volume, liters/min	Energy expenditure METS
Before sleep	70.0 + 10.7	17.0 + 4.8	9.1 + 3.3	1.8 + 0.8
After sleep	58.0 + 10.5	16.0 + 5.1	9.4 + 3.4	1.2 + 0.7
After the 3rd h of physical work	77.0 + 10.7	21.0 + 5.2*	15.5 + 3.0*	3.3 + 0.7

* Significance of differences: $p < 0.05$.

postulated that CIVD would decrease the overall ability to survive in the cold. However, our data suggest that when the skiers were warm and in no danger of hypothermia, their CIVD response was enhanced. When there was general cooling, the CIVD response was reduced. enabling them to survive better in the cold; it seems that there is a local adaptation to protect against cold injury of the extremities, and also a general insulative-hypothermic adaptation to a general cold stress.

Reactions of Cardiorespiratory and Central Nervous Systems to Soviet Chamber Simulations

The cardiovascular data indicated only a moderate psychophysiological reaction, with a limited impact upon the heart rate, respiratory minute volume and energy expenditure (table 8.3). Cardiac dynamics were evaluated as in chapter 6, looking at the variance of the resting heart rate. However, the several indices (table 8.4) suggested only an insignificant stressing of the cardiorespiratory system and its autonomic regulation.

The functional state of the central nervous system has a major impact upon the extent and direction of adjustments to a stressful environment, whether simulated or "real". Chamber observations were based upon differentiated self-appraisal indicators (table 8.5) and simple and complex visuomotor reaction times. Analysis of the latter was analogous to that adopted for the electrocardiogram, emphasis being placed not only upon mean values, but also upon the variance of data (table 8.6).

No significant changes in the self-appraisal indicators were seen. The initial scores generally showed a high level of nonspecific and modelled operator's skills. Differences between the mean and mode for the simple visuomotor reaction time suggested there may have been some decline in the motivation and/or emotional attitude of the subjects during testing.

Table 8.4. Variance of the resting electrocardiogram during climatic chamber simulation of the ski trek in Moscow (mean ± SD)

Stage	Duration of cardiac cycle, ms	Variance, ms	Coefficient of variation, %
Before sleep	888 ± 126	68 ± 28	7.6 ± 2.3
After sleep	1,061 ± 181	51 ± 18	5.1 ± 0.8
After the 3rd h of physical work	790 ± 133	80 ± 28	6.6 ± 2.2

No signifiicant ($p < 0.05$) differences were identified.

Table 8.5. Differentiated self-appraisal indicators (SAM test) of the participants ($n = 12$) during chamber simulation of the ski trek (mean ± SD)

Stage	State	Activity	Mood
Before sleep	5.3 ± 0.5	4.9 ± 0.8	5.5 ± 0.7
After sleep	5.4 ± 0.7	4.9 ± 0.8	5.5 ± 0.8
After the 3rd h of physical work	5.7 ± 10.0	5.2 ± 0.9	5.5 ± 0.8

No significant ($p < 0.05$) differences were identified.

Table 8.6. Indices of simple and complex (color discrimination) reaction times of participants during climatic chamber simulation of the trek in Moscow

Participant	Simple visuomotor reactions				
	mean ms	variance ms	coefficient of variation %	mode ms	amplitude % ms
D.I. Shparo	243	64	26	207	29
F.F. Konyukhov	243	38	15	204	23
V.I. Shishkarev	290	87	30	225	18
M.G. Malakhov	237	54	22	222	26
A.V. Meknikov	255	54	20	245	34
A.A. Belyaev	17	34	15	204	39
A.P. Fedyakov	249	65	29	203	25
Yu. I. Khmelevsky	219	38	17	207	47
M. Buxton	248	57	23	203	22
L. Dexter	271	57	21	253	28
R. Weber	305	53	17	284	19
M ± SD	245 ± 24	54 ± 18	22 ± 6	215 ± 15	28 ± 4

Mode of cardiac cycle, ms	Amplitude %	Range of variation, ms	Stress index arbitrary units
883 ± 144	19.0 ± 5.5	398 ± 230	44 ± 46
1,066 ± 174	22.0 ± 4.8	244 ± 70	47 ± 24
786 ± 123	23.0 ± 10.8	252 ± 96	74 ± 59

This assumption was supported by an increased variation span for the simple visuomotor reaction time. The chamber simulation apparently imposed no stress upon the central nervous system. The integral indices of the variance in sensometry (the functional level of the system, FLS, the reaction stability, RS, and the level of functional possibilities, LFP) were judged to be at a medium to a high normal level relative to the classification developed by Loskutova [1975].

The highest functional potential (a rapid mean reaction time, an insignificant difference between mean and mode, and the highest FLS, RS and LFP scores) were shown by A.A.B. and Y.I.K. The lowest functional

	Complex reactions				
range of variation, ms	FLS	RS	LFP	mean ms	variance ms
518	4.6	1.8	3.3	359	53
199	4.1	1.0	2.4	329	67
424	4.1	1.0	2.4	275	44
344	4.2	1.4	2.8	338	57
363	4.6	2.1	3.4	238	44
0280	4.6	2.1	3.6	321	51
383	4.4	1.4	2.9	315	118
281	5.0	2.7	4.2	293	65
291	4.4	1.3	2.8	266	39
466	4.2	1.6	2.9	293	43
275	3.5	0.6	1.5	333	43
349 ± 98	4.5 ± 0.3	1.5 ± 0.1	3.1 ± 0.6	306 ± 45	64 ± 19

potential was seen in V.I.S. and R.W., although it was thought that the slow reactions of V.I.S. reflected poor motivation (for further comments on motivational aspects of the psychophysiological tests, see chapter 9).

The complex reaction test involved color discrimination. Data were treated in similar fashion. The most successful subjects (smallest number of incorrect responses in the shortest time) were B.V.M., M.B. and L.D., while the least successful subjects were M.G.M., A.A.B. and V.P.L.

Relation to Mission Success

Cannonical correlations were used to relate the Soviet psychophysiological data to mission success, as discussed further in chapter 9. The FLS and LFP were each positively correlated with both self-appraisal (SAM) and expert co-appraisal, while there were also negative correlations between expert appraisal and measures of test variance. In essence, the more successful the subject was during the initial chamber simulation, the higher the self- and co-appraisals at the conclusion of the actual mission.

There was no significant relationship between the simple visuomotor reaction time as such and mission success.

9 Psychic Adaptation of Participants[1]

V.S. Koscheyev, M.A. Lartzev, A. Rode, M. Malakhov

General Considerations

Overall Effects of Stress

The term adaptation, as used by Russian psychologists, is somewhat at variance with the genetic connotation given to this word in the literature on acclimatization to unusual environments; where the text seems to permit this, adjustment or acclimatization has thus been substituted by the editors.

The psychic reactions of the individual, and the interaction of such reactions with physiological processes are of primary importance in successful acclimatization to and work within unusual environments. Disturbances of psychic adjustment play a major role in development of both the 'polar stress syndrome' [Kaznacheev, 1974] and the nonspecific 'psycho-emotional stress syndrome' that is commonly reported at high latitudes. Both syndromes result from an increased level of anxiety, with its psychophysiological and biochemical correlates. Disturbances of the acclimatization process manifest themselves in these syndromes, with destabilization of neuropsychic functions, a narrowing of the potential range of helpful psychic reactions, a reduction in the accuracy of performance of complex sensorimotor tasks, and an increased risk of various psychosomatic diseases [Kaznacheev, 1974; Berezin, 1980; Androvna et al., 1982].

Stresses Associated with High Latitudes

The stresses experienced at high latitudes can be classified into natural and psychosocial problems. Natural difficulties include the extremely cold climate, disturbances in the normal hours of daylight (photoperiodicity), possible geophysical influences from the altered electromagnetic field, intense solar irradiation during the arctic summer, abrupt changes in the barometric pressure, high wind velocities and whiteouts [Androvna et al., 1982; Anastazi, 1982]. The altered hours of daylight upset the normal stereotypes of the brain, leading to profound disturbances in sleep patterns, a decreased performance efficiency and a general deterioration of condition.

---

[1] Freely adapted from the Russian translation by R.J.S.

Such disturbances seem particularly acute during the period of continuous polar night. However, natural factors are unlikely to explain interindividual differences in the processes of psychic and physiologic acclimatization, since the natural features of the north tend to be uniformly extreme and threatening to health and life for all inhabitants of such territory [Chubinnskij, 1965].

Many authors thus emphasize the key role that is played by personal psychological and psychosocial influences [Kaznacheev, 1974], both in inducing the polar stress syndrome and in allowing acclimatization to this environment [Vasilevskij et al., 1978; Matusov, 1979; Berezin et al., 1980]. The average person is confronted by a breakup in the normal stable stereotypes of life, with a loss of established interpersonal relations. Commonly, there is isolation and relative sensory deprivation, with separation from those nearest and dearest to the individual. The choice of partners for interpersonal communication becomes extremely limited. Finally, the sociocultural norms of the northern habitat are often very different from those previously experienced in a more southerly environment.

Migration to high latitudes has complex and often contradictory motives. The move to the north may be made under the force of circumstances. Factors influencing such migration include a yearning for exotic surroundings, material incentives such as 'free' housing or high salaries, job requirements, difficulties in interpersonal communication experienced elsewhere (in effect, a running away from undesired contacts), difficulties in professional adjustment in larger cities, professional failure, a search for new spheres of activity, and a search for an environment that places less constraints upon personal behavior. A number of these factors can influence the behavior not only of the long-term immigrant, but also of the volunteer for an arctic expedition.

Psychic Adaptations to Stress

The patterns of psychic adjustment associated with the ski-trek have been viewed within the general framework previously formulated by Berezin [1980]. According to this concept, 'psychic adaptation is a process ensuring the optimal correlation of the personality and the environment in the course of implementing characteristic human activity'. In conformity with Berezin's [1980] view, the better the psychic adjustment, the more effectively the individual will satisfy personal needs and attain desired goals, fitting behavior to the requirements of the natural and the social environment, and avoiding intrapsychic and interpersonal conflicts.

In instances when goal attainment stresses psychological abilities 'close to the limit of surpassing them, it appears possible to talk about emotional stress' [Basowitz, 1955]. Threat is a necessary element of emotional stress, and whether it will be perceived or not depends on a combination of personality traits [Curtis, 1985; Kaznacheev, 1974; Gotes and Banham, 1961; Borchgrevink, 1971; Miroshnikov, 1971; Gubachev et al., 1976] and prior experience. No combination of factors or situations is equally stressful to all individuals; much depends upon their psychological and psychophysiological peculiarities [Basowitz, 1955].

The subjective reaction to the blocking of some urgent need, frustration, is no less important to psychic adjustment than is stress, particularly in the conditions of the far north where a wide range of biological and social needs often cannot be met.

A third concept relevant to mechanisms of adjustment is that of searching activity [Arshavskij and Rotenberg, 1984]. The mere absence of negative emotions is in some cases insufficient to maintain psychic or even somatic well-being. Stress can also arise from monotony, where the sole unfavorable feature is the permanence of the situation.

Finally, the experience of negative emotions may promote a breakdown, with the development of pathological processes that activate the capacities of the organism for acclimatization [Arshavskij and Rotenberg, 1984]. Such a possibility is supported by the occasional dramatic decrease in the incidence of diseases, infections and psychosomatic diseases at times of substantial and acute emotional stress, even if the stress is caused by negative experiences such as wars and natural disasters. In contrast, protracted periods of stress posing a constant threat to personal welfare are associated with a decreased resistance to diseases.

Implications for Ski Trek

The investigators sought to establish consistent relationships between individual peculiarities of personality, the psychic state as observed and recorded during the ski trek, and the success or otherwise of skiing activity while crossing the polar ice-cap. A further issue of some interest was the psychological evaluation of interactions between representatives of two very different cultures and sociopolitical systems (the Soviet and Canadian participants) during an expedition that involved some 90 days of autonomous activity under extreme conditions.

Potential outcomes of the research included an improvement in techniques for forecasting the likely functional state and performance of individuals undertaking autonomous expeditions at high latitudes, and the responses of teams of professional athletes undertaking comparable feats.

An attempt was thus made to assess the initial characteristics, psychic state and personality of each of the participants, to related the dynamics of psychic state during the expedition to this initial data, and to relate both sets of characteristics to mission success as measured by expert appraisal and objective indicators.

Methods for the Study of Psychic Adjustments

Clinical Approach

Complementary clinical/descriptive and experimental/psychological approaches were adopted [Anastazi, 1982]. The traditional clinical approach is descriptive and generalizing. It explores in a unique manner the diverse relationships between psyche and soma, conscious and unconscious, and biological and social events. Individual characteristics are analyzed, incorporating into the general picture phenomena that are unique yet inaccessible to experimental observation. Particular emphasis is placed upon the identification and assessment of anomalies arising under the extreme conditions of the expedition, with an evaluation of their impact upon behavior and activity patterns during and following the mission [Berezin, 1980].

Use of Standard Questionnaires

The experimental/psychological approach used primarily standard psychodiagnostic tools to provide objective individual and group assessments of current psychic state and personality, relating such data to external events ranging from severe changes in the natural environment to differences in the protective equipment available to individual participants [Anastazi, 1982]. Note was taken of both the intensity and success of activity during the trek, and the objective data was also related to the clinical observations. Formal questionnaires have the practical advantage that they can be completed by the entire team of participants simultaneously, an important consideration given the limited time available for experimental measurements. Moreover, while the presence of a psychologist is desirable when administering such test instruments, it is possible to collect information during the trek in the absence of the principal investigator, scoring responses after return to a southern laboratory.

The selected standard instruments were the Minnesota Multiphasic Personality Inventory (MMPI) and Cattell's 16PF test. Each of these tests allows simultaneous evaluation of a number of traits, an important consideration in a situation where there is no a priori indication of those characteristics that are most likely to be affected by the adverse environment. Since some of the team spoke only Russian, and some only English, it was necessary to select procedures that were available in both languages [Anastazi, 1982]. National normalized standards for these two tests permitted the direct comparison of scores for Canadian and Soviet participants.

The MMPI test includes special scales to evaluate the reliability of responses. Combinations of abnormal responses also permit the identification of particular syndromes such as the neurotic triad that can be correlated with clinical observations.

Cattell's 16PF allows an examination of specific traits in the emotional and intellectual spheres, together with factors related to interpersonal communication, teamwork, and reactions to frustration.

Subjective Assessments

Subjective assessments reflect objective processes in the consciousness or the sensations of a subject. Scores on such assessments reflect the attitude to self, the psychic state, and the current attitude towards other persons, processes and objects in the immediate environment. Subjective assessment is influenced by a variety of psychological mechanisms, at every stage of psychic activity from perception to the construction of complex conceptions. Such mechanisms can distort subjective assessment, particularly when the information that is analyzed comes into conflict with conscious and subconscious notions of the self (the so-called self-concept). Subjective ratings are also influenced by intelligence, by the emotional connotation of a particular situation, and by the motivation of the individual.

Subjective assessment is particularly important during an expedition of the type described here, since objective criteria of success are extremely limited, and laboratory scientists do not have the opportunity to observe behavior and activity patterns directly while the trek is proceeding. The self-appraisal of mood (SAM) rated three items – general state, activity and mood. Questionnaires were prepared in Russian and English. The main advantage of the SAM test was its brevity. Responses were supplemented by a detailed questionnaire that evaluated on a series of 5-point Likert scales physical fitness, the level of preparedness for the journey, and some psychosocial characteristics. Each participant evaluated all other team members, allowing both co-appraisal and subsequent expert reevaluation of the reports. The expert appraisal was compared with self- and co-appraisal, and was finally used to rank the various team members.

Other Test Procedures

A sociometric questionnaire aimed at identifying the most active members of the team, those with whom conversation was most sought, and those who contributed most to the needs of the traverse.

Raven's progressive matrices [Anastazi, 1982] were used to test the ability of team members to solve problems of graphic logic within a specified time limit. The test had the advantage of being relatively rapid to administer, and independent of linguistic ability. Somewhat surprisingly, the Soviet investigators reported that scores were related to the effectiveness of social adjustment, and also helped to predict both professional performance and deviancy in social behavior.

Lüscher's color test [Anastazi, 1982] provided further data on individual personality traits. This test is based on color preferences, subjects ranking in order of preference eight standard colors. Again, the test was nonverbal, and thus equally applicable to Soviet and Canadian participants.

The experimental procedures were complemented by extended interviews, which yielded additional information on personality traits, general mood state, character of social behavior and professional activity.

Experimental Plan

The initial examination (MMPI, 16PF, Raven's Progressive Matrices, color test and interview) was conducted in Moscow during daylight hours. Soviet participants also completed the expert and self-appraisal questionnaires.

During the traverse, it was planned to complete further MMPI and 16PF questionnaires, to complete further expert and self-appraisal questionnaires, and to keep detailed psychological diaries relating to each 'leg' of the itinerary. Detailed instructions were given to assist the participants in undertaking these duties, although the Soviet participants already had extended experience in the use of these instruments. Unfortunately, problems of motivation led to failure to complete some of the evaluations.

Clinical psychodiagnostic surveys were planned at Cape Arkiticheskij, just before the trek began, and at each of the subsequent mandatory medical examinations held at two week intervals during the expedition.

Final observations were planned for the immediate end of the trek, at Cape Columbia, and in Ottawa.

Initial Status of Participants

For ethical reasons, detailed scores for individual participants are not discussed. The average MMPI and 16PF scores for the Soviet team members are shown in figures 9.1 and 9.2. Scores showed no major departures from normality, with the exception of the B scale of the 16PF (which relates to intelligence). Other deviations from the anticipated values

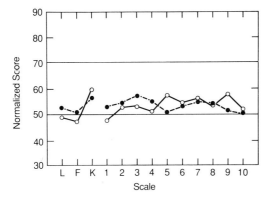

Fig. 9.1. Average MMPI scores for Soviet participants before (●) and after (○) the expedition.

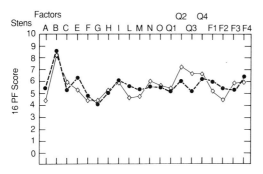

Fig. 9.2. Average 16PF scores of Soviet participants before (●) and after (◇) the expedition.

of 50 and 5.5 on the two tests are expressed as a selection of descriptions of behavior; in the final analysis these must be matched against direct observations of the individuals concerned. Commonly, the description given below outruns the objective underpinnings of the questionnaire scores. The interpretation of the Soviet psychologists reflects this dualistic approach, and leads to a detailed if speculative account of the personality of those electing to participate in the trek.

The Soviet observers suggested that the average initial scores reflected some overanxiety about health, a lack of self-assurance and an increased attention to various negative signals. Moreover, team members tried to anticipate and prepare for even insignificant and unlikely contingencies, with a striving to avoid failures and disappointments rather than to focus upon success at any price. Where there seemed a risk of failure, participants experienced a need to give up. Situations with an unpredictable outcome, a fast pace and a lack of orderliness that precluded advanced planning were seen as stressful. When making decisions, Soviet participants strove to analyze the greatest possible number of options, and experienced difficulty in making the best choice among these options. In consequence, there was anxiety, tension, preoccupation about specific situations, and a moderate deterioration of activity and mood state. At the same time, the group was characterized by a restrained tendency to realize their drives and needs regardless of the immediate situation, to some extent disregarding existing standards, social taboos, and rules of behavior in achieving their goals. Team members were pragmatic, concerned about their own interests, somewhat egotistic, unstable and oriented towards personal achievement rather than group activity. A desire to avoid possible failures led to restraint, preventing the open manifestation of antisocial tendencies. However, such adverse behaviors were liable to occur when it was possible to shift the blame for specific frustrations to others, for instance by attempting to induce feelings of anxiety and guilt in the individuals concerned. Increased irritability, hot tempers and a lack of restraint were also noted. Such characteristics became manifest in response to slips, blunders and failings of associates, which provided a pretext for demonstrations of irritation.

The Soviet participants also exhibited some features of dominance, an inclination to assume leadership and demonstrate independence. They showed perseverance in attaining goals that had been set if they could see feasible ways of attaining them, they stood up actively for their rights, and they acted in accordance with the principle that 'ends justify the means'. At the same time they showed features of sensibility, impressionability and enhanced emotionality. There was a yearning for new impressions, a guidance by esthetic criteria, imagination and artistic perceptions of the world, combined with a distaste for coarse people and 'crude' activity. The group tended to deny the existence of conflicts and difficulties encountered in their interpersonal communications and in the control of their own behavior, striving to produce a favorable impression while winning the support and sympathy of their associates. As a result, they commonly gave the impression of being sensible, well-disposed and sociable people with a broad range of interests. A great experience in interpersonal contacts and refusal to admit existing difficulties made them highly enterprising and capable of finding an acceptable policy while realising their needs. They made adequate social contact with other team members, but showed no active need for communication. They also had a highly developed logical type of thinking, with abilities for analysis and generalization (table 9.1).

Table 9.1. Raven's test scores for participants

Participant	Number Completed	Percentage of errors
D.I. Shparo	46	0
V.I. Shishkarev	57	0
M.G. Malakhov	45	2
Yu.I. Khmelevsky	51	5
V.P. Ledenev	43	7
A.P. Fedyakov	30	44
A.V. Melnikov	46	7
F.F. Konyukhov	32	33
A.A. Belyaev	56	2
L. Dexter	44	2
R. Weber	43	7
M. Buxton	48	0
K. Holloway	56	7

Self-appraisal of the Soviet participants yielded somewhat conflicting data. On the one hand, there was an inclination towards analysis and a certain degree of diffidence, but on the other hand there was a displacement from consciousness of negative information that could have deflated their self-appraisal. These internal contradictions apparently reflected the desire to win sympathy and recognition from their peers, coupled with doubts about their worthiness for such recognition.

Comparison of Soviet Team with Participants in Previous Expeditions

A comparison of findings on the present Soviet team members with the psychological data obtained on members of a 1984 expedition showed a number of important differences. In 1988, the average scores for the group showed greater anxiety and tension, more doubt as to the team's abilities to realize objectives, and a consequent fall in mood state and performance efficiency. A desire for conflict-free ways to attain goals, gentleness and tractability had apparently been replaced by aggressiveness, an inclination to assume leadership roles and poor manageability. Conformity and a desire to meet social norms and rules of behavior had been replaced by irritability, a lack of restraint, and a tendency to protest any encroachment upon personal interests. The ability to sympathize with others and emotionality (especially in communication) were less marked. The social distance between team members was increased, and interests were less unified.

Self-control was weaker, and the group more easily became disorganised in critical situations. The level of aspirations was increased, but participants were less vigorous in attaining their goals. The planning of behavior and activity was less precise, and concern about public reputation was lessened.

Direct interviews and observation of Soviet team members in Moscow and Dikson largely confirmed conclusions drawn from the experimental data. The Soviet participants looked worried, tense, and expressed their concerns about preparatory measures. They were plainly disturbed by the final composition of the team, which had not been resolved until a very few days before beginning the trek. There also appeared to be conflicting relations and poor psychological compatibility between some members of the team. Relationships between the group were generally formal and business-like, without any expressions of sympathy at an emotional level of communication. Signs of displeasure with each other, irritation and the censuring of real or alleged blunders were expressed practically without restraint, even in the presence of outsiders. The impression was formed that the participants were connected only by their activities in preparing for the trek. In essence, they were functioning as a production team rather than as an informal group.

The attitude of the team towards the scientific aspects of the program was unfortunately somewhat negative. It was asserted not infrequently that the political goals of the expedition (cooperation between the USSR and Canada, and the demonstration of arctic sovereignty) far outweighed any scientific objectives, and that the latter were in the final analysis not very necessary to success of the mission.

With respect to the psychological tests, most team members formally expressed a readiness to undertake the required tests, but in practice their execution was postponed or forgotten, with claims of fatigue or shortness of available time. Some participants were explicitly displeased, and objected to the experimental studies, particularly the psychodiagnostic tests. It is hypothesized that such reactions stemmed from a reluctance of potential participants to reveal their psychic state when the results of such analysis might adversely affect their likelihood of inclusion in the team. On the other hand, the greatest lack of discipline in performing the desired experimental tasks was shown by individuals whose participation in the trek was beyond question. These individuals apparently saw the implementation of the scientific program, and especially the time-consuming psychodiagnostic tests as an additional burden which was hampering the completion of the expedition's primary mission. Given these circumstances, difficulty was encountered in completing all the desired psychological measurements. Specifically, it was necessary to complete some of the initial tests in Dikson rather than in Moscow, as was intended.

Comparisons with data from the 1984 expedition suggest that as the interrelations between the team became more distant, emotionally cold, and less proper, their motivation to undertake the scientific program diminished, and relationships with the individuals conducting the examinations became less trusting. The style of the expedition's leadership also became more authoritarian.

Characteristics of Canadian Participants

The Canadian participants were analyzed by a combination of English language MMPI, 16PF, and the observations of the Soviet investigators (fig. 9.3, 9.4). While some apparent differences in mood state and personality were noted between the Canadian and Soviet team members, it is less clear how far these were attributable to barriers of communication between the two linguistic groups, differences in the size of the Canadian and Soviet teams, and differences in their sociocultural backgrounds.

The Canadian group included a physician and a clergyman. Two of the Canadians were also champion cross-country skiers. In the view of the Soviet psychologists, the Canadian team exhibited a positive background mood, moderate activity, and an adequately high efficiency of performance. They were confident, optimistic, spontaneous and sociable. They made decisions with ease, were free from internal conflicts and contradictions, and enjoyed change. They switched easily from one type of activity to another, and easily changed their points of view. Their self-appraisals were not inflated, and they showed no tendency to interpret the actions of their associates as intended hostility. Their way of thinking was commonplace, and when making decisions they did not try to be original but chose traditional optimal solutions. They were gentle, and showed no inclination to be domineering, nor did they demonstrate tractability or manageability. They tended to be demonstrative, and strove to win the recognition and sympathy of their associates, underscoring their activity, optimism and self-assurance. At the same time, they could be hot-tempered and unrestrained, and protested against infringements upon their interests. They cared more about their own interests than public duty, and were somewhat egotistic, pragmatic and business-like. They also tried to restrain emotional manifestations, and on occasion were cold and distant. Despite their apparent sociability, they experienced no need for emotional contacts with their associates. They were characterized by emotional maturity, permanence, realism of their assessments and adequate self-control, and they did not become disorganized in critical situations. They were vigorous and enterprising, actively pursued their intended policy and strove to achieve

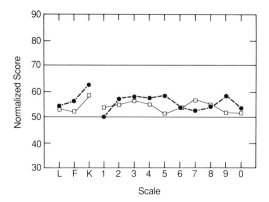

Fig. 9.3. Average MMPI scores of Soviet (□) and Canadian (●) participants before the expedition.

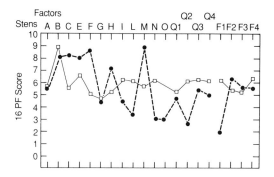

Fig. 9.4. Average 16PF scores of Soviet (□) and Canadian (●) participants before the expedition.

the goals that they had set. They were not afraid of conflicts. They tended to shift the blame for frustrations upon their associates. They were traditionalist in neither lifestyle nor behavior, and took no pains to appear ordinary. Rather, they could be regarded as eccentric or even peculiar. However, they demonstrated some commonsense features, being rational, sober-minded and practical. They did not try to be competitive – they were easy-going rather than remote, straightforward rather than smart or crafty, unpretentious and natural, did not conceal their assessments, showed no inclination to scheme, and valued straightforwardness and sincerity. They needed support, advice and approval from their peers, and tended to work as a group. Their levels of aspirations and the intensity of their needs were not particularly high. Finally, they demonstrated a logical type of thinking and the ability to analyze and make generalizations (table. 9.2).

Table 9.2. Ranking of participants by summated expert appraisals before and after the trek

Before the expedition			After the expedition		
Rank by expert appraisal	participant	summated expert appraisal score	Rank by expert appraisal	participant	summated expert appraisal score
1	V.I. Shishkarev	4.76	1	V.I. Shiskarev	4.81
2	R. Weber	4.70	2	M.G. Malakhov	4.78
3	M.G. Malakhov	4.58	3	V.P. Ledenev	4.63
4	V.P. Ledenev	4.56	4	R. Weber	4.50
5	A.V. Melnikov	4.53	5	A.A. Belyaev	4.51
6	D.I. Shparo	4.	6	M. Buxton	4.36
7	L. Dexter	4.27	7	Yu.I. Khmelevsky	4.33
8	F.F. Konyukhov	4.04	8	D.I. Shparo	4.33
9	A.P. Fedyakov	3.87	9	K. Hallowoy	4.21
10	Yu.I. Khmelevsky	3.82	10	A.V. Melnikov	4.07
11	A.A. Belyaev	3.81	11	L. Dexter	3.94
			12	A.P. Fedyakov	3.92
			13	F.F. Konyukhov	3.51

Comparison of Soviet and Canadian Participants

The initial data showed some significant differences between average scores for the Soviet and Canadian teams. The Canadian participants were more active, more sociable and spontaneous, and free of internal conflicts. The levels of anxiety, tension and concern about their status was also lower than that of the Soviet group. The Canadians controlled their behavior more easily, seemed emotionally more mature and became less disorganized in situations of stress. They were also more open and trusting, less inclined to interpret the actions of their colleagues as manifestations of hostility, and had a much greater need for peer support and group activity. Unlike their Soviet counterparts, they were eccentric, original and less traditionalist. Their levels of aspirations and intensity of needs were lower than those of the Soviet group.

In direct contacts before the start of the trek, the Canadian participants gave the impression of being sociable and uninhibited people, any restraint being attributable to contact with a different culture and ignorance of some of the rules of behavior in the new environment. They entered easily into conversation, answered questions amiably and expressed optimism when assessing the prospects of the expedition. They looked quiet and undisturbed, and showed no signs of anxiety, worry or disorganized behavior. With one exception, the request to undergo psychodiagnostic testing was accepted without a negative reaction. One member of the Canadian team argued that he was very busy with organization of the crossing, and further did not understand the goals of the psychological testing. As a group, the Canadians felt no embarrassment in speaking about their general situation, their families and their hobbies. They also showed no negative attitudes towards the Soviet participants or the culture and customs of the USSR, avoiding political discussions.

On the other hand, the Canadians were displeased by the request to complete the psychodiagnostic tests while en route, complaining about the extra weight of the psychodiagnostic diaries and reminding the Soviets that they had brought portable tape recorders to register their impressions during the trek.

Data Obtained During and Following the Trek

During the trek, the Canadians were unwilling to complete the psychological measurements, so most pieces of data are available only for the Soviet participants (fig. 9.5–9.8).

First Drop

At the time of the first parachute 'drop' of supplies, the state of the participants was characterized by an enhanced peculiarity of thinking, and by a shaping of unusual attitudes and original approaches to the solution of traditional problems. Difficulties had appeared in the assessment of the social situation, and notions about a behavioral style necessary for mutual understanding had become blurred. Because of the peculiarities of thinking, the range of stimuli causing emotional reactions narrowed, probably as a form of psychological defence. Contacts among the participants became more superficial, in an attempt to lessen interpersonal conflicts and to reassess the significance of such conflicts. The participants seemed to become more apathetic and indifferent to what was happening around them. At the same time, there was a growing interest in physical health, an increasing number of physical complaints, and a desire to limit activity to a minimum because of the deterioration in general health. Overall behavior became rigid and stereotypic, with a focus upon maintaining personal physical health. Physical problems, and concern about them were apparently used as a pretext to explain in rational terms the alienation and estrangement from associ-

9   Psychic Adaptations of Participants

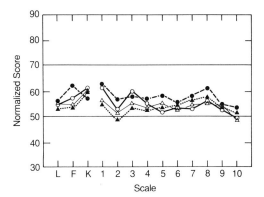

Fig. 9.5. Averaged MMPI indicators for Soviet participants at different stages in the expedition. ● = 1st stage; ○ = 3rd stage; ▲ = 4th stage; △ = 5th stage.

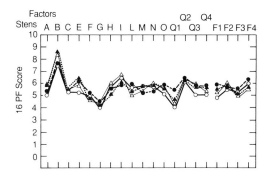

Fig. 9.6. Average 16PF indicators of the group of Soviet participants of the expedition at different stages of the crossing. ● = 1st stage; ○ = 3rd stage; ▲ = 4th stage; △ = 5th stage.

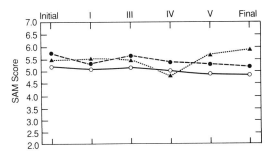

Fig. 9.7. Average SAM indicators of Soviet participants at different stages of the expedition. ● = General state; ○ = activeness; ▲ = mood. Data for initial state at drops 1, 3, 4 and 5, and after completion of the trek.

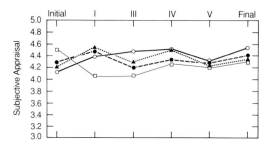

Fig. 9.8. Average indicators of subjective assessment of individual traits and activity of participants. ● = Expert appraisal of the Soviet participants; ○ = self-appraisal of the Soviet participants; ▲ = co-appraisal of the Soviet participants; □ = expert appraisal of the Canadian participants. Data for initial state at drops 1, 3, 4 and 5, and after completion of the trek.

ates. At this stage, there was a tendency to overrate the importance of difficulties and failures.

The participants had also become more anxious and tense, with doubts about their endurance. At the same time, there was a marked tendency to block aggression, with a diminished desire for dominance and competition, and an enhanced attention to emotional nuances in contacts with other team members. This last change helped to lessen tensions between members of the expedition.

On the whole, deviations in psychological data from population norms were greatest at this stage of the expedition, implying a maximal stress upon the mechanisms of psychic adjustment, and a nadir of effective functioning.

Findings at Third Drop

By the time of the third parachute 'drop' was reached, the psychic state of the participants was characterized by much lower levels of anxiety and tension, with greater self-confidence and a disappearance of the previous tendency to magnify difficulties. However, the concern about physical health was unabated, and there was even an increased tendency to use physical complaints in order to escape unwanted duties.

Other features at this stage were dominance, aggressiveness, independence and an inclination to leadership. The participants became more sociable, less distant and less alienated. They made decisions with greater ease, and no longer had difficulty in choosing the best options.

Findings at Fourth and Fifth Drops

The psychic state progressively optimized over the later legs of the journey, corresponding to the fourth and fifth supply 'drops', with most of the test scores returning towards population norms. Anxiety and tension were further alleviated, the features of hypochondria were less evident, and activity of the sympathetic nervous system was apparently optimized. The team members were now less demonstrative, cared less about external effects of their behavior, and ceased to use physical complaints to influence the actions of their associates. In general, behavior became more 'proper', with a tendency to conformity. Impulsivity, lack of restraint and thoughtlessness were less characteristic, the background mood improved, and aggressiveness, independence and a tendency for leadership were less expressed.

Findings after Completion of the Trek

Final observations were made in Ottawa, shortly after completion of the trek. As compared to observations made during the traverse, there was a sharp decline in concern about personal health, the number of physical complaints was reduced, and the general condition was improved. The participants were now more active and sociable. Their behavior was better controlled, and they were less irritable and impulsive. Adherence to social norms and rules of behavior had increased, the planning of activity and emotional control were optimized, and there was a reduced probability of disorganization in stressful circumstances.

This final stage of the mission was marked by a further decrease in dominance and aggressiveness, and attention was increasingly focussed upon the emotional nuances of interpersonal relations. At the same time, the need for group interaction and peer support diminished to its lowest level. However, manifestations of anxiety and tension were greater than during the final stages of the journey.

Comparing findings with the initial data, the main differences were a higher level of activity, a greater independence, a lesser demonstrativity, a lesser desire to attract attention, and a lesser aggressivity and desire for dominance. Behavior was more carefully controlled after the traverse, and the participants were less impulsive and irritable. There was also less tendency to infer manifestations of hostility from the actions of other team members, although the participants also became more cool and distant from each other. Much less attention was paid to personal health, and there were fewer physical complaints, reflecting a parallel sympathetic stabilization. After the journey, the participants were less demonstrative, paid less attention to external factors, and ceased to use physical complaints as a lever against their associates.

Subjective Appraisals

Methodology

All participants were asked to complete questionnaires assessing their own performance and that of their peers (fig. 9.8). At the initial assessment, impressions were also related to experience gained in previous arctic expeditions. Members of the Soviet team found themselves unable to make assessments on two new Canadian skiers with whom they had no previous experience as travelling companions, while the Canadian team refused to fill out these questionnaires on the grounds that they lacked sufficient experience in communicating with their Soviet counterparts. During the crossing, all participants were asked to appraise themselves and other members of the team, limiting their impressions to specific segments of the journey. Finally, they were asked to complete similar questionnaires related to the entire traverse.

Initial Data

When first undertaken, the expert appraisal of the Soviet participants was greater than their co-appraisal, which in turn was greater than self-appraisal. The low scores at self-appraisal reflect anxiety, tension and the general lack of confidence of the Soviet group at the beginning of the trek. Expert appraisals of two Canadians exceeded their personal appraisals, forecasting very successful participation in the mission.

Assessments during the Trek

As the journey proceeded, scores for all appraisals increased. At the first re-supply 'drop', there was agreement between expert and co-appraisal for the Soviet team, although self-appraisal was still at a lower level than either of these measures. The correspondence between expert and peer appraisal may reflect a mobilization of the respondents and a consolidation of grouping among the Soviet team in the face of a strong challenge. At the same time, the expert appraisal of the Canadian participants was fairly low, likely reflecting the perceived lack of training of the Canadians for a crossing of this type, possible boosted by some prejudice on the part of the Soviet group against what they regarded as poorly trained partners.

Subjective assessments were not completed at the second 'drop'. By the third 'drop', the previous pattern had become inverted in the Soviet participants, with self-appraisal being highest, co-appraisal next, and expert appraisal lowest. Possibly, this was connected to fatigue and weakness of the team, with a decline of functional possibilities, and less precise assessment of personal qualities. Enhanced self-appraisal might also be underpinned by successful completion of the most difficult part of the crossing, with a progressive disappearance of anxiety and depression. However, the main change was a decrease of peer and expert appraisal rather than an increase of self-appraisal. Because of these changes, the expert appraisal of the Soviet team moved downward towards the continuing low scores for the Canadian team members.

At the fourth 'drop', the trends discerned at the third 'drop' persisted. Self-appraisal of the Soviet group again exceeded expert appraisal, while the rating of the experts now quite closely matched the perceptions of the Canadian participants. The growing parity of all appraisals may reflect a smoothing of the influence of conflicts upon scoring at this stage.

By the time of the fifth 'drop', all subjective assessments were virtually coincident (fig. 9.8), which may reflect a consolidation of the team and a reduction of contradictions among its members. At the same time, a decrease in the absolute values of all three assessments points to fatigue and weakness among the participants, with a growing exactitude of judgments.

Final Data

The final assessments that were made in Canada showed an insignificant increment in all appraisals, with a growing closeness of expert ratings for the Soviet and Canadian members of the expedition.

Relationship of Appraisals to Mission Performance

The efficiency of performance was rated and ranked by the expert panel in a manner intended to reflect fitness for activity, using data obtained before and after the trek (table 9.2). The same individuals remained the perceived leaders before and after completion of the trek, and the hierarchy of Soviet participants was strictly maintained, with only one team member advancing in ranking. However, the Canadian participant who was initially ranked quite highly was placed lower after completion of the trek. Three Soviets and one Canadian were finally ranked as 'outsiders', with an undesirably poor performance.

In order to explore further relationships between mission success and the psychological data, three highly ranked individuals (V.I.S., V.P.L. and M.G.M.) were compared with the three lowest ranked Soviets (A.V.M., A.P.F. and F.F.K.). Owing to the small sample size, it was not possible to assign statistical significance to the perceived differences between these two subgroups (fig. 9.9–9.14).

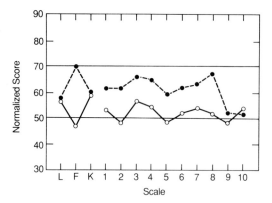

Fig. 9.9. Average initial MMPI score for successful and unsuccessful participants. ○ = High success; ● = low success.

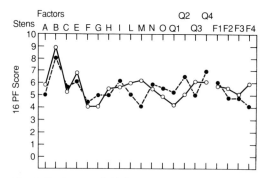

Fig. 9.10. Average initial 16PF scores for successful and unsuccessful participants. ○ = High success; ● = low success.

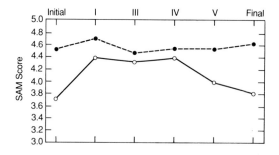

Fig. 9.11. Subjective assessments of successful and unsuccessful participants at different stages of the expedition (expert assessments). ● = High success; ○ = low success.

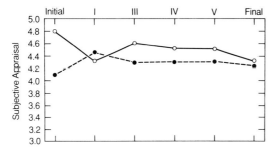

Fig. 9.12. Subjective assessments of successful and unsuccessful participants at different stages of the expedition (self-appraisal). ● = High success; ○ = low success.

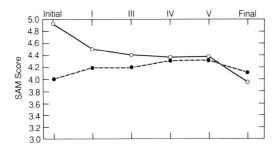

Fig. 9.13. Subjective assessments of successful and unsuccessful participants at different stages of the expedition (co-appraisal). ● = High success; ○ = low success.

Profile of Successful Participants

At the outset of the expedition, the 'best' members of the expedition were marked by moderately expressed traits of demonstrativity, a desire to win support and sympathy, and conformity to social norms. They showed no tendency to overrate their difficulties, preferring to project the best possible impression and to emphasize their harmonious relationship with others. Levels of anxiety and stress were relatively low, and they were sufficiently self-confident. At the same time, they carefully thought through all options and were sufficiently independent, with an ability to think logically, to analyze and to generalize, easily making decisions with few logical mistakes when time was short.

Based upon data obtained in 1982–1984 (the 'Komsomalskaya Pravda' expedition), the psychic adjustments in all of this group were stable and effective. Anomalies of personality were not detected either initially or during the traverse. Further, the personality traits of all three participants were close to those for the average population.

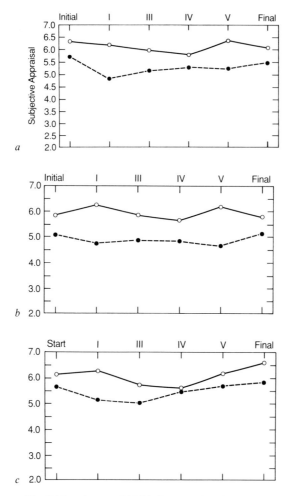

Fig. 9.14. *a* Average SAM indicators of Soviet participants with high and low success at different stages of the expedition (general state). ● = High success; ○ = low success. *b* Average SAM indicators of Soviet participants with high and low success at different stages of the expedition (activity). ● = High success; ○ = low success. *c* Average SAM indicators of Soviet participants with high and low success at different stages of the expedition (mood). ● = High success; ○ = low success.

Profile of Least Successful Participants

The psychological profile of the least successful participants showed a number of traits that differentiated the subgroup sharply from the most successful participants. They showed much higher levels of anxiety and tension, with a lack of self-confidence. Activity and mood state were

9 Psychic Adaptations of Participants

depressed. They were more impulsive, more irritable, and had difficulty in either controlling their behavior or in forecasting the negative implications of their actions. Peculiarities of thinking led to inadequate assessment of situations and wrong decisions. Communication with other members of the team was also ineffective, and their motives were not always comprehensible. When making decisions, they tended to consider many options, and had difficulty in choosing between them.

The least successful participants were demonstrative, yet when trying to win support and sympathy, they used inappropriate means. Their behavior was thus perceived as contradictory, and sometimes as incomprehensible. They also tended to overrate difficulties, and to draw their perceived problems to the attention of others.

Pronounced personality anomalies were identified in one of the three least successful participants in the period before the trek commenced. Two of this group showed a pronounced decline in their ability to think logically, and made many mistakes in solving logical problems under time pressures. Data from 1982–1983 confirmed that the psychic adjustment of two of the three 'worst' participants was unstable and not sufficiently effective, with a tendency to manifest anomalies of personality when under severe physical and/or emotional stress.

The average personality scores for the unsuccessful subgroup differed markedly from the 'best' participants, also showing much larger deviations from population norms. On the other hand, self-appraisal yielded higher scores in the least successful group.

Observations during the Traverse

The psychological characteristics of the 'best' participants remained stable and within population norms throughout the traverse, confirming the stable and effective nature of their psychic adjustment. However, the psychic state of the 'worst' participants in some instances suggested a high level of tension and ineffective psychic adjustment. Personality anomalies were apparently aggravated in two of the latter group as the traverse continued. The third member of the 'unsuccessful' subgroup (who had been thought to have no personality anomalies before the expedition began) tended to close himself off from psychological examination, while demonstrating nonpathological peculiarities such as aggressivity, dominance, and an inclination to leadership. In the Soviet view, such tendencies would lead to stress and disturbances in psychic adjustment.

Observations on Completion of Trek

When the trek was completed, the 'best' participants remained sufficiently effective (fig. 9.15, 9.16). Relative to the period immediately before

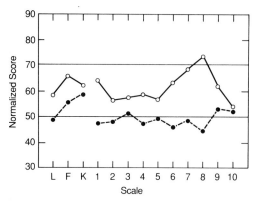

Fig. 9.15. Average MMPI scores for participants with high and low success after completion of the expedition. ● = High success; ○ = low success.

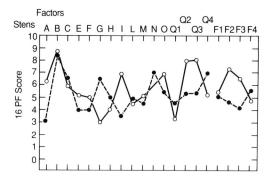

Fig. 9.16. Average 16PF scores for participants with high and low success after completing the trek. ● = High success; ○ = low success.

beginning the trek, they exhibited a decline in demonstrativity, and were less worried about the impression that they produced. Their interpersonal contacts became more superficial and less emotional, while their interests became more earthy and utilitarian. At the same time, the 'best' participants became more restrained and correct in their behavior. They were less irritable and hot-tempered, less inclined to conflicts, and less likely to interpret the actions of others as hostile, while their emotional control improved. Their background mood became more positive, but activity and efficiency of performance decreased as they became more relaxed. They also paid less attention to their general physical condition, and their thinking became quite commonplace.

The psychic adjustment of the 'worst' participants remained at a low level of effectiveness after the trek was completed (figs. 9.15, 9.16). Pronounced peculiarities of thinking, anxiety, hypochondriasis, lack of self-assurance and worry about their state of general health were still evident. At the same time, activity was enhanced, and because this was combined with anxiety, it led to chaotic behavior. Interpersonal contacts became less discriminating. Demonstrativity was reduced, as was the need to attract the attention of others, and there was no longer a tendency to use physical complaints for this purpose. Impulsiveness, irritability and a lack of restraint were all toned down. At the same time, behavior became more egotistic, and was directed to the satisfaction of personal interests, with a decreased need for group interaction and support. All three members of the 'worst' subgroup showed persistent anomalies of personality after completion of the trek, suggesting that completion of their mission had not resulted in any substantial decline of psychic stress or optimization of their psychic state.

Comparison of Best and Worst Participants

Direct comparison of the 'best' and 'worst' participants (fig. 9.15, 9.16) showed that the 'best' participants were less anxious, were free of hypochondriasis, were less predisposed to seek an excessive number of options, were able to discriminate between primary and secondary issues, and thus found decision making easier. Traditional patterns of thinking helped the 'best' participants to assess various situations more adequately and to establish interpersonal contacts with greater ease. Their behavior was perceived as more comprehensible, and their emotional reactions were seen as more appropriate to their causes.

The 'worst' participants were more vulnerable, sensitive, suspicious, unfriendly and prone to conflict. They felt that they had been undeservedly hurt and underestimated. At the same time, they were more emotional, responsive and strove to understand the motives behind the actions of others. They were also more concerned about their general health, made a greater number of physical complaints, and blamed interpersonal difficulties, estrangement and alienation upon their physical problems. After completion of the mission, the 'worst' participants were more prone to concentrate on their own interests, questions of prestige, and the social implications of their participation in the expedition.

The subjective assessments also showed differences between the two groups at all stages of the trek (figs. 9.12–9.14). Expert appraisal was higher for the 'best' participants at every evaluation. Over the first four drops, the differences between the two subgroups diminished, suggesting that there was some adjustment on the part of the least successful participants, but the differences became larger again at the fifth drop, and there was a further decrease in appraisals of the 'worst' participants after the trek was completed.

Before the trek began, the self-appraisal was higher in the 'worst' than in the 'best' participants (fig. 9.11), but this over-appraisal diminished by the first drop. Possibly, the stresses and difficulties encountered to this point led to a more realistic self-appraisal. However, by the third drop, the 'worst' subjects were again rating themselves more highly than the 'best'. Thus, it may be concluded that the 'worst' participants lacked an adequate perception of their abilities as members of the team. All three aspects of the self-assessment (general state, activity and mood) were also rated more highly by the 'worst' participants throughout the mission (fig. 9.14) suggesting either unrealistic self-assessment or a wish to dissimulate personal difficulties.

Before the expedition began, peer appraisal rated the 'worst' participants more highly than the 'best'; however, the scores assigned to the 'worst' fell progressively as the trek continued, while scores for the 'best' progressively improved.

Underlying Motivations

The initiative for the expedition came from members of the Soviet team, and in general they played a dominant role in making the necessary arrangements. Primary leadership was also from a Soviet team member (D. S.). All of the Soviet participants were volunteers who had come together as a result of previous Arctic expeditions. They jointly devised an appropriate route for the crossing, and an appropriate sharing of duties during the trek.

Most of the participants saw the trek as a form of recreation, although subsequent governmental support gave the expedition official status, with political and educational goals, a potential for substantial logistic support, and a need for professionalization of the endeavor. Nevertheless, it remained essentially an amateur undertaking, with no rigidly defined responsibilities except with respect to duties that were essential to the crossing. The prestige of participation was enhanced not only by the uniqueness of the route, but also by substantial media coverage of the trek. Participants saw a prospect of recognition, awards, and competition from other potential members of the trek. The prestige of the trek was further enhanced in 1988 when the Soviet team were joined by Canadian representatives, and the expedition was granted the status of an international political activity.

An informal selection of team members was an essential feature. The weeding out of unsuitable participants was essentially based upon 'survival of the fittest', and a demonstrated psychological compatibility with members of the core group. Because of such self-selection, the motivation of the group was much higher than that of many occupational groups who are required to work in high latitudes, and the findings cannot be generalized to an occupational setting.

Canadian participants were selected in a somewhat similar fashion. The main criteria were a high level of physical fitness and psychological compatibility with their Soviet colleagues. The process of natural selection undoubtedly eliminated Soviet potential participants who were less resistant to the stresses of extreme environments, as well as many who were poorly motivated to overcome such factors. Two of the Canadians had no prior experience of this type of expedition, although they had lived in high latitudes. Two of the Canadians also lacked training in long-distance skiing and the carrying of the heavy backpack needed for autonomous existence in this type of environment. It thus seems likely that the stresses imposed upon these individuals were more severe than those encountered by the remainder of the team.

Many aspects of the northern environment that are popularly assumed to be extreme induced no emotional stress in most of the participants. Likewise, the inevitable frustrations of such an expedition, including a blocking of biological and social needs, were generally well accepted as a necessary feature of the trek. Indeed, many of the participants apparently had an active desire for a situation where they were exposed to stress and privations, and this undoubtedly reduced the extent of their emotional reactions.

Overall Analysis of Behavior

Behavior was determined more by the stable characteristics of individual personalities than by immediate social and sociopsychological factors. In particular, there was a high motivation to avoid failure and to achieve success.

The desire for success encouraged a completion of the arduous daily schedule of physical activity, and even gave a positive emotional coloring to this activity. The expectation of a favorable outcome led to optimism, a high level of activity, and a desire to be at the center of attention, with a wish to gain sympathy, support, attention and recognition from colleagues. Simultaneously, this motivation was reinforced by internal needs to overcome difficulties, to conquer doubts about personal strength and to demonstrate steadfastness. Failure avoidance was a further motivation.

The Canadian participants seemed characterized particularly by achievement motivation, as shown by the absence of anxiety and internal conflicts, a willingness to take risks, and a firm belief in the correctness of their plans and actions. In contrast, the Soviet team members seemed more concerned with the avoidance of failure, as shown by anxiety, internal conflicts, lack of self-assurance, and the contradictory motives that had induced them to join the expedition. Their activities were undertaken with a desire to consider all the assessments and expectations of their colleagues, carefully weighing all decisions and rejecting those fraught with a risk of failure because of a fear of censure. In some cases, failure avoidance led to a lack of activity, or even an active curtailing of competition inside and outside the group.

One aspect of motivation was probably 'searching activity' [Arshavskij and Rotenberg, 1984]. A transpolar expedition may be viewed as a form of behavior that changes a life situation that (consciously or subconsciously) the individual finds dissatisfying. In essence, there is a search for an acceptable form of activity, with a wish to escape from the stresses of everyday life. In the view of the Soviet psychologists, such searching activity enhances the activity of mechanisms of psychic adjustment and blocks pathological processes, even if there are negative emotions aroused by the need to overcome specific difficulties. In support of this view, the psychological data generally show a progressive optimization of psychic state under conditions which most people would have regarded as intolerable. The systematic persistence with which many team members strove to subject themselves to the extreme environment of the far north lends further credence to this idea.

Within this broad theme, a number of more specific motives may be identified: (1) a search for novelty and exotic environments, with an escape from the monotony of everyday life; (2) a search for public attention to compensate for a perceived or unperceived dissatisfaction with current social status, and (3) an attempt to correct a lack of self-confidence through the development of physical and mental powers, and a wish to confirm personal ability to withstand privation.

Interviews and observation suggested important differences in motivation between the Soviet and the Canadian participants. The Soviets were anxious to rid themselves of deep intrapsychic conflicts through some special form of activity. However, for the Canadians, the most appealing feature of the mission was its uniqueness, its athletic challenge, and the element of competition with other team members. In general, the motives of the Canadian participants seemed less ambiguous and less contradictory than those of their Soviet peers. However, the differences in motivations between the representatives of the two nations also led to some conflicts and hampered coordination of the team's activities.

Group Dynamics

While in Moscow and Dikson, group dynamics were characterized by a high level of anxiety and tension, with a desire to examine all options very carefully. It is likely that there was a high level of stress at this period, because of organizational problems and uncertainties by many of the Soviets as to whether they would be included in the trek. A high level of anxiety and diffidence was undoubtedly promoted by the authoritarian style of the expedition's management, and by some fuzziness in definition of the criteria for the selection of participants.

Lower levels of anxiety, uninhibited behavior and self-assurance of the Canadian participants reflected both a differing culture, and the certainty that they would participate in the mission. A favorable psychic state was also encouraged by their youth, a high level of physical fitness, and sound physical health.

The stressing of the psychic mechanisms during the first stage of the traverse seems attributable to very severe weather conditions. Moreover, during this phase even the experienced participants had to reacclimatize to the extremes of cold that were encountered. The strain upon the autonomic nervous system manifested itself in hypochodriacal tendencies. There was also a shift of values towards a dominant concern for physical health, physical fitness, and an all-round conservation of effort. Against the background of such tendencies, interpersonal conflicts became less significant and caused no dramatic or energy-consuming reactions. Conflicts were alleviated by promoting the idea that survival in such circumstances was dependent upon group activity. For the Canadian participants, the first stage involved a learning of the skills of survival and working in the high arctic. Their refusal to perform the psychodiagnostic tests during the traverse may have been motivated in part by their extremely pronounced desire to conserve their physical and mental powers.

Irritability and impulsiveness during the early part of the crossing may be viewed as a consequence of physical weakness and insufficient adjustment of psychic mechanisms. The total failure of the team to complete the psychodiagnostic tests at the time of the second 'drop' provides indirect evidence of the weakness and extreme stress upon team members.

After the third 'drop', the psychic state improved, probably due to the passing of the most severe weather conditions. Protective psychological mechanisms that persisted at this stage included hypochondriasis and demonstrativity. Anxiety and manifest stress diminished, and the control of behavior improved. However, an inflated self-appraisal at this stage indicated a lack of awareness of their limitations by some team members.

Further optimization of psychic state was inferred at the fourth and fifth 'drops'. However, final observations in Ottawa revealed some recurrence of anxiety, tension and concern about personal status. Possibly, such reactions reflect a return of the conflicts associated with a 'civilized' life, qualitatively very different from the difficulties encountered during the traverse, and (for the Soviet participants) quite different from the life known in the USSR. There may also have been a surfacing of some emotions that had been suppressed during the traverse. Nevertheless, most participants were in a more favorable psychic state at the end than at the beginning of the expedition.

Identification of Successful Participants

The psychic adaptation of the 'best' participants was effective and stable, this subgroup showing personality characteristics close to population mean values. Neither the social changes associated with departure and completion of the mission, nor the natural hazards and sociopsychological conflicts encountered en route induced any unfavorable changes in their psychic state. They showed a high level of intelligence on the Raven tests of logical ability, and in tasks requiring analysis and generalization. Moreover, this same intellect helped them in solving technical problems and selecting optimal behavior. Their self-appraisals remained very stable, and were either deflated, or close to the ratings assigned by the experts.

The unsuccessful participants were characterized by ineffective, unstable and deteriorating psychic adjustments over the course of the expedition, with anomalies of personality that were either generally manifest or emerged in stressful circumstances. These features suggested an enhanced probability of deviant behavior and failures. The Raven matrices showed their difficulty in logical thinking when under time constraints, with a lack of ability to analyze and generalize. Self-appraisal was inflated.

Four members of the Soviet team demonstrated anomalies of personality and intellectual function. Three of these were the least successful participants. The fourth person was initially assessed as the worst, but he had risen to the mid-point of rankings by the conclusion of the expedition. Nevertheless, some features of his behavior were not adaptive, especially during interpersonal contacts. The Canadian participant who initially showed some anomalies of personality, in contrast, was a moderately successful participant.

Self-appraisal was not significantly correlated with expert appraisal ($r = 0.18$), but there was a statistically significant correlation ($r = 0.40$) between success and features of the psychological questionnaires. Positive attributes included factor B (logical thinking), factor E (perseverence, purposefulness, dominance), and factor Q_4 (ambition, level of aspirations, intensity of needs). Negative correlations were seen for MMPI scale 2 (level of anxiety, depressive tendency), scale 3 (demonstrativity, propensity to external influences), scale 6 (inflated self-appraisal, sense of hostility in the environment), and Cattell factor I (sensitivity, dreaminess, dislike of crude relationships). Moreover, all MMPI scores were negatively correlated with success, irrespective of the magnitude of the correlation.

Despite the demonstration of these relationships, there did not appear to be one optimal set of characteristics guaranteeing successful completion of the expedition. Similar degrees of success could be achieved through varying combinations of physical fitness, efficiency and motivation. In particular, some team members were able to discover new ways of achieving their objectives through the use of their intellect. Thus, only major disturbances of personality should be regarded as unfavorable characteristics, although the demonstration of psychic disturbances and pathological reactions at the initial examinations or during simulation of peak loads indicate a somewhat poorer prognosis. In the MMPI test, note should be taken of standardized scores of over 70 (two SDs greater than population norms), the most unfavorable findings being high scores on scales, 2, 3, 4, and 6. The probability of deviant behavior and mission failure apparently rises in proportion to the height of scores on these scales. On the 16PF test, caution should be shown if a subject demonstrates low scores on factors B, C, E, B, Q_1 and Q_4 and a rise in factor I. On the Raven matrices, more significance should be attached to a substantial number of mistakes of logic than to the number of items completed in the allocated time. In terms of the self-appraisal, a score that exceeds peer and expert appraisals and inflated SAM scores is again an adverse finding.

The ideal participant seems effective and stable, with an absence of personality anomalies indicative of stress upon the mechanisms of acclimatization. If anomalies are present, the prognosis is more favorable if the psychic state improves progressively over the course of observation. The intragroup sociopsychological microclimate is favored by conformity, a striving to meet social and ethical norms of behavior, an ability to take account of the interests and opinions of associates, a need for emotional contact and support, yet independence in decision making. Aggressivity, dominance and a propensity to assume leadership should not be too marked. When making decisions, the ideal participant should review each of several options, but not have difficulty in deciding between them. Necessary perseverance and purposefulness should not lead to an inflated self-appraisal or a feeling that colleagues are insufficiently attentive or hostile. It is important to be able to compare self-appraisal realistically with peer appraisals. The desire for conflict should be low, but not to the point of excluding independent thinking. Thought patterns should be sufficiently concrete as to facilitate communication, mutual understanding and group activity. The background mood should be positive and active, without superficiality or inconsistency. Contacts with other team members should be made easily, and at a sufficient depth to satisfy the mutual need for communication without being excessively intense or intrusive. Emotionality must be sufficient to empathize with colleagues while avoiding an excessive increment of group tensions. Ideally, a supportive reaction to a partner should be shown if the difficulty or failure is warranted by the circumstances. A rational explanation should be advanced, allowing for a noncontradictory self-concept and rejection of a succession of fruitless attempts to satisfy a particular need.

Intelligence is an important attribute, particularly an ability for rapid, logical thinking, analysis and generalization, with a minimum of errors. Intellect also seems important to social behavior, and an ideal team would have little differences in intellectual function between participants.

Finally, appropriate sources of motivation are critical. The optimal motives behind participation are communication, service to others and participation in scientific research, while a predominance of materialistic incentives, a need for publicity and other egotistical motives seem most undesirable.

Conclusions

The studies of psychic adjustment to the stresses of the ski-trek were based on the concepts of Berezin [1980], where the body seeks an optimal match of personality with environment in the course of performing a given activity. Another underlying concept was to seek a unified system of social, scientific, engineering, ergonomic and medical measures that would allow complete professional adjustment of workers to the total environment of northern latitudes. Throughout, the stress was on professional success, with less analysis of personal satisfactions.

The success of an expedition such as the present ski-trek is determined by a combination of a high level of function in those recruited, a well-designed system of selection, adequate physical and psychological preparation, stable personalities and a high level of individual and group motivation, with careful planning of organizational, engineering and medical needs.

In the current expedition, the participants in general were characterized by a high initial functional status, with insignificant stress upon their regulatory mechanisms. The three team members with an inadequate adjustment were identified as early as the preparatory stages of the mission.

The stress upon regulatory mechanisms was increased during the trek, due to a combination of cold discomfort, considerable physical loads, fatigue, and interpersonal conflicts, particularly those arising between Soviet and Canadian participants. Although statistically insignificant, there were signs suggestive of inadequate adjustment and deregulation of the body defences against psychic stress, with such physical manifestations as a decrease in working capacity and clinically significant changes in the ECG suggestive of right heart overload. Interindividual differences in these responses were strongly influenced by each participant's personality and current psychic state.

Appendix A

Peer-rating used during traverse

Date .... Stage of crossing ....

Please answer the questions presented below, having in mind the results of the last 10-day leg of the crossing. Choose 2 or 3 participants of the expedition the most fitting to the question asked and list them in the descending order by the quality assessed (the first is the best).

1. Whose contribution to the success of the last leg of the crossing is the greatest in your opinion?

    1.   2.   3.

2. Who is the most vigorous and active when skiing?

    1.   2.   3.

3. Who is the most vigorous and active when working at stopping places?

    1.   2.   3.

4. If there was some hard and uninteresting work who was the most willing to handle it?

    1.   2.   3.

5. Who was ready (or suggested) to give you help without waiting to be asked?

    1.   2.   3.

6. Name those members of the expedition with whom you are most prepared to cooperate in such expeditions in future as based on the facts of the last leg of the crossing.

    1.   2.   3.

Appendix B

Self-appraisal used during traverse

Full name.. Date ....

I feel well	3 2 1 0 1 2 3	I feel bad
I feel strong	3 2 1 0 1 2 3	I feel weak
Passive	3 2 1 0 1 2 3	Active
Sedentary	3 2 1 0 1 2 3	Lively
Merry	3 2 1 0 1 2 3	Sad
Good spirits	3 2 1 0 1 2 3	Low spirits
Capable of working	3 2 1 0 1 2 3	Feeling jaded
Full of strength	3 2 1 0 1 2 3	Collapsed
Slow	3 2 1 0 1 2 3	Fast
Inactive	3 2 1 0 1 2 3	Energetic
Happy	3 2 1 0 1 2 3	Unhappy
Cheerful	3 2 1 0 1 2 3	Gloomy
Strained	3 2 1 0 1 2 3	Feel slack
Healthy	3 2 1 0 1 2 3	Sick
Apathetic	3 2 1 0 1 2 3	Carried away
Indifferent	3 2 1 0 1 2 3	Anxious
Enthusiastic	3 2 1 0 1 2 3	Cheerless
Joyful	3 2 1 0 1 2 3	Mournful
Relaxed	3 2 1 0 1 2 3	Tired
Renewed	3 2 1 0 1 2 3	Exhausted
Sleepy	3 2 1 0 1 2 3	Excited
Desire to rest	3 2 1 0 1 2 3	Desire to work
Calm	3 2 1 0 1 2 3	Worried
Optimistic	3 2 1 0 1 2 3	Pessimistic
Capable of endurance	3 2 1 0 1 2 3	Weary
Hale and hearty	3 2 1 0 1 2 3	Listless
Grasp slowly	3 2 1 0 1 2 3	Grasp quickly
Absent-minded	3 2 1 0 1 2 3	Attentive
Full of hopes	3 2 1 0 1 2 3	Disappointed
Self-satisfied	3 2 1 0 1 2 3	Displeased

9 Psychic Adaptations of Participants

Dear Sir,
Please give an estimate of some peculiarities of all expedition members, using a 5-number evaluation system; 5 means the best (give an estimate of yourself)

Date of completion of the form ....

No.	Full name	Development level estimate physical / mental / moral	Estimate of the general state of health	Estimate (forecast) of successful participation in the expedition	Relative estimate of the participation in the expedition	Estimate of a man as a companion	Personal relation to a companion	General contribution to the expedition activities
1.								
2.								
3.								
4.								
5.								
6.								
7.								
8.								
9.								
10.								
11.								
12.								
13.								
14.								
15.								

Full name of person completing in the form ....

10 Status of Selected Hormones and Biologically Active Compounds

R.A. Tigranian, N.F. Kalita, N.A. Davydova,
B.R. Dorokhova, M.G. Malakhov, A.G. Melkonian,
A.S. Roganov, I.D. Stalnaya, T.I. Chernikhovskaya,
T.N. Shumilina

Introduction

Hormone concentrations are normally assayed in whole blood, plasma, or serum. While changes in the observed concentrations are commonly interpreted as evidence of altered secretion, such changes can also reflect altered metabolism, excretion or binding [Shephard, 1983]. In studies of the ski trek, an attempt has been made to evaluate these complexities by looking at patterns of blood and urinary concentrations, and to test ratios of trophic hormones against their products, interpreting data in terms of favorable and unfavorable adaptations to the various stresses of the traverse (metabolic, thermal and psychological).

A large fraction of many of the hormones found in the blood is normally bound to transport proteins, and is thus biologically inactive. It is important from a functional point of view to distinguish free from total blood concentrations, and again the data has drawn this distinction with respect to certain hormones such as the thyroid complex.

It is well recognized that blood sampling itself can be a traumatic experience, particularly when attempted under field conditions, and that the blood levels of many hormones are somewhat susceptible to acute stresses in addition to more long-term patterns of response [Shephard and Sidney, 1975]. Moreover, the hormones form a complex 'orchestra' of symphonic proportions, with complicated interrelated fugues and feedback mechanisms, so that the unravelling of acute or chronic responses to any type of stress is extremely complex.

Earlier studies [Kalita et al., 1984, 1987; Krylov and Tigranian, 1984, 1986: Krylov et al., 1985; Tigranian et al., 1985, 1988a, b] have suggested that residence in the high north is accompanied by significant adjustments of neuroendocrine regulation. Hormones and neuromediators play an important role in acclimatization to extreme conditions, and it was thus thought useful to investigate blood and urine levels of a wide range of hormones, biologically active compounds, neuromediators and metabolic products before, during and after completion of the transpolar ski trek. The responses observed were intended to reflect the adjustments of the participants to the combined impact of low temperatures and heavy physical loads, although some have questioned how far baseline data obtained in Moscow were representative of the initial status of the team, considering the psychological stresses experienced as preparations were made to transport food, equipment and personnel for Moscow to the departure point in the Severnaya Zemlya Archipelago.

It proved possible to repeat some of the baseline data at Dikson, just before commencement of the trek, and this information is regarded as more reliable from the viewpoint of the stabilization of the emotional state of the team members.

Methodology

Blood samples were collected in Moscow (18 days before the trek), in Dikson (11 days before), on days 12, 29, 55 and 74 of the 91-day traverse, and in Ottawa on day 3 after completion of the mission. 24-hour specimens of urine were collected on the same days as the blood specimens (with the exception of day 29). A field centrifuge was used to separate plasma and serum. The blood and urine samples were frozen, and after completion of the expedition all specimens were flown to Moscow in dry ice, for detailed biochemical analysis.

Standard techniques of radioimmunoassay were used to determine the content of hormones, neurotransmitters, neuropeptides and other biologically active compounds, using commercially available reagent kits from: CIS, France, for adrenocorticotropin (ACTH), cortisol, prolactin, luteotropin (LH), follicle-stimulating hormone (FSH), aldosterone, renin, β-endorphin, somatostatin, somatomedin C, parathormone, calcitonin, testosterone, and substance P; from Mallinckrodt, FRG, for reverse triiodothyronine (rT_3); from Hoechst, FRG, for growth hormone (GH), thyroid-stimulating hormone (TSH), C-peptide; from Serono, Italy, for glucagon; from Amersham, UK, for free triiodothyronine (fT_3), free thyroxine (fT_4); from Bühlmann, Switzerland, for antidiuretic hormone (ADH), angiotensin II; from DRG, USA, for serotonin, neurotensin, prostaglandin E (PGE), angiotensin-converting enzyme (ACE); from CSSR for 3′,5-adenosine monophosphate (cAMP), 3′,5-guanyl monophosphate (cGMP); from Hungary for thromboxane, 6-keto-$PGF_{1\alpha}$, and from USSR for insulin, estriol, estradiol.

Radioactivity was measured using a RIA-gamma 1271 automatic gamma counting system (LKB, Sweden) and a Rackbeta 1217 liquid scintillation system (LKB, Sweden).

The adrenaline, noradrenaline (NA), dopamine (DA) and DOPA content of the blood and urine were determined fluorimetrically [Euler and Lishajko, 1959]. The level of metanephrine (MN) and normetanephrine (NMN) in the urine was assayed by the method of Matlina et al. [1974]. Urinary concentrations of vanillylmandelic acid (VMA) and homovanillic acid (HVA) were determined by the thin-layer chromatography method of Dluskaya et al. [1967]. Blood and urinary concentrations of serotonin, histamine (HM), histidine (HD), tryptophan, 5-hydroxytryptophan (5-HOT) and 5-hydroxyindoleacetic acid (5-HIAA) were measured as proposed by Gerasimova [1977]. The functional activity of sympatheticoadrenal, serotoninergic and histaminergic systems were evaluated by the calculation of coefficients expressing the relative rates of synthesis and metabolism of catecholamines, serotonin and HD [Bolshakova, 1982].

The blood content of a number of metabolites (triglycerides, cholesterol, total lipids) was estimated enzymospectrophotometrically or colorimetrically, using standard Boehringer kits (FRG) [for further details, see chapter 11]. Blood glucose levels were determined by the glucose oxidase method, using a Beckman (USA) glucoanalyzer.

Blood and urine concentrations of sodium and potassium were determined by flame photometry, while osmolality was estimated by a cryoscopic method (Osmette 2007 osmometer, Precision Systems, USA).

Methods of statistical analysis included Duncan's multiple range test [Duncan, 1955], Student's t test [Urbach, 1975] and Wilcoxon's nonparametric criterion [Gubler and Genkin, 1973].

Sympathoadrenal, Serotoninergic and Histaminergic Systems

General Considerations

The sympathoadrenal system produces two important catecholamines adrenaline and NA. The sympathetic nerve endings secrete NA plus small amounts of adrenaline, while the output of the adrenal medulla is some 75% adrenaline and 25% NA. In our present context, these compounds have important effects upon the heart and blood vessels (increase in myocardial contractility, heart rate and cutaneous vasoconstriction). Through β_1 receptors, they also stimulate lipolysis, while through the β_2 receptors (more responsive to adrenaline than to NA) glycolysis is enhanced.

Previous studies on rats [Strabrovskii and Korovin, 1972] have shown increased blood and urinary levels of catecholamines in response to cold, with decreased adrenaline reserves and increased NA in the adrenals and myocardium. Such changes were considered as evidence for the release of adrenaline and NA from the adrenals, with a compensatory increase in NA production. Other animal studies have confirmed these findings [Kvetnansky et al., 1971; Bostic et al., 1979]. The rate of ^{3}H-NA mobilization from the rat hypothalamus is increased during the first 6 h of cold exposure [Simmonds, 1971], again indicating a cold-induced increase in catecholamine release. Human observations confirm increased blood levels of NA in the cold [Jessen, 1980], while the concentrations of adrenaline have either increased or remained unchanged [Wilderson et al., 1974; Hiramatsu et al., 1984; O'Malley et al., 1984; Scriven et al., 1984; Leppäluoto et al., 1988]. Field observations have shown substantial increases in blood and urinary catecholamine levels in skiers under arctic conditions [Malakhov et al., 1982; Kalita et al., 1984, 1987; Krylov and Tigranian, 1984, 1986; Krylov et al, 1985; Tigranian et al., 1985, 1988a, b].

The basic response of the circulatory system to cold is a redistribution of blood flow, aimed at decreasing perfusion of peripheral tissues with an increase in flow to vital and heat-producing tissues [Barbarash and Dvurechenskaya, 1986]. NA and serotonin are both involved in these responses. NA also has a pronounced effect on thermogenesis [Strabrovsky and Korovin, 1971; Pastukhov and Khaskin, 1979], including any nonshivering thermogenesis [Sellers, 1957; Schonbaum et al., 1963] and it thus contributes substantially to thermal homeostasis during cold acclimatization [Yakimenko, 1982]. Finally, NA is heavily implicated in the change from carbohydrate to lipid metabolism that occurs during prolonged work [Marshak, 1965; Meerson, 1972; Kulinsky and Plotnikov 1982]. The involvement of adrenaline and the sympathetic nervous system in vascular aspects of thermal regulation are well known [Slonim, 1969; Barbarash and Dvurechenskaya., 1986]. Serotonin also contributes to heat conservation [Gromova, 1966].

Serotonin accumulates in platelets through an active transport process. It has marked vasoconstrictor properties, and is released during tissue injury (such as may occur with both severe cold exposure and prolonged exercise). Histamine levels are also increased by both tissue injury and prolonged exercise.

The serotoninergic system is further involved in regulating the general level of central nervous system activity, including motor activity, emotional behavior, memory and training processes. Emotional stress intensifies the modulating effects of both serotoninergic and noradrenergic brain systems [Naumenko and Popova, 1975].

Changes during the Expedition

During and after the trek, blood concentrations of adrenaline and NA significantly exceeded initial values, the maximal increase being observed

Table 10.1. Content of neurotransmitters and some metabolites in the blood of participants (mean ± SD)

Location	Adrenaline μg/l	Noradrenaline μg/l	Serotonin ng/ml
Moscow	–	–	–
Dikson	0.70 ± 0.03	0.87 ± 0.01	109.9 ± 8.7
Day 12	2.87 ± 0.24^b	2.33 ± 0.17^b	163.4 ± 18.5^b
Day 29	3.30 ± 0.26^b	3.63 ± 0.23^b	157.8 ± 18.3^b
Day 55	2.62 ± 0.12^b	2.40 ± 0.09^b	134.1 ± 12.2
Day 74	1.49 ± 0.07^b	1.73 ± 0.04^b	58.7 ± 6.8^b
Ottawa	0.96 ± 0.08^b	1.05 ± 0.05^b	115.0 ± 9.3

Days 12, 29, 55 and 74 are described as stages 1, 2, 3 and 4 of the traverse, respectively, in the text.
^{a,b}Significant differences relative to the tests conducted in Moscow (a) and Dikson (b).

Table 10.2. Urinary excretion of catecholamines, their precursors and conjugates (mean ± SD)

Location	Adrenaline μg/24 h	Noradrenaline μg/24 h	Metanephrine μg/24 h
Moscow	11.0 ± 0.6	33.7 ± 0.7	156.2 ± 2.5
Dikson	14.4 ± 1.0^a	35.7 ± 1.2	168.0 ± 1.4^a
Day 12	22.9 ± 2.6^{a,b}	45.6 ± 3.0^{a,b}	183.2 ± 4.2^{a,b}
Day 74	26.6 ± 1.5^{a,b}	54.2 ± 2.4^{a,b}	216.3 ± 5.7^{a,b}
Ottawa	19.3 ± 1.4^{a,b}	49.0 ± 2.2^{a,b}	180.8 ± 1.3^{a,b}

^{a,b}Significant differences relative to the tests conducted in Moscow (a) and Dikson (b).

Table 10.3. Excretion of vanillylmandelic acid (VMA), histamine, serotonin and their metabolites (mean ± SD)

Location	VMA mg/24 h	Histamine μmol/24 h	Histidine μmol/24 h
Moscow	3.26 ± 0.1	0.72 ± 0.1	591.0 ± 46.1
Dikson	3.92 ± 0.2^a	1.10 ± 0.1^a	632.8 ± 31.0
Day 12	5.07 ± 0.4^{a,b}	1.62 ± 0.2^{a,b}	685.0 ± 32.6
Day 74	8.03 ± 0.5^{a,b}	1.98 ± 0.1^{a,b}	761.9 ± 44.9^{a,b}
Ottawa	7.53 ± 0.4^{a,b}	1.12 ± 0.1^a	726.7 ± 40.2^a

^{a,b}Significant differences relative to the tests conducted in Moscow (a) and Dikson (b).

Glucose mmol/l	Cholesterol mmol/l	Triglycerides mmol/l	Total lipids g/l
5.63 ± 0.07	6.80 ± 0.25	1.79 ± 0.06	9.81 ± 0.9
5.06 ± 0.10[a]	7.11 ± 0.26	1.70 ± 0.11	10.5 ± 0.7
4.67 ± 0.04[a,b]	6.72 ± 0.22	1.62 ± 0.08	11.7 ± 1.4
4.23 ± 0.14[a,b]	6.6 ± 0.25	1.58 ± 0.08	14.0 ± 0.6[a,b]
4.62 ± 0.20[a]	6.91 ± 0.32	1.39 ± 0.03[a,b]	12.9 ± .0[a]
4.23 ± 0.13[a,b]	7.40 ± 0.40	2.70 ± 0.13[a,b]	13.8 ± 0.2[a,b]
4.66 ± 0.05[a,b]	7.31 ± 0.31	1.57 ± 0.06[a]	10.9 ± 0.5

Normetanephrine μg/24 h	Dopamine μg/24 h	DOPA μg/24 h	HVA mg/24 h
127.3 ± 2.6	313.5 ± 13.8	32.7 ± 1.2	2.53 ± .0.1
141.6 ± 4.2[a]	352.6 ± 11.2[a]	37.1 ± 0.6[a]	2.83 ± 0.2
164.7 ± 5.6[a,b]	404.9 ± 17.1[a,b]	48.2 ± 2.6[a,b]	4.04 ± 0.3[a,b]
188.0 ± 3.6[a,b]	461.7 ± 17.2	56.3 ± 1.4[a,b]	6.21 ± 0.4[a,b]
160.5 ± 5.4[a,b]	406.5 ± 15.1[a,b]	44.5 ± 2.0[a,b]	5.38 ± 0.3[a,b]

Serotonin μmol/24 h	Tryptophan μmol/24 h	5-HOT μmol/24 h	5-HIAA μmol/24 h
0.64 ± 0.1	144.0 ± 6.5	1.43 ± 0.1	15.3 ± 0.9
1.01 ± 0.1[a]	139.3 ± 5.3	1.28 ± 0.1	20.8 ± 1.2[9]
1.51 ± 0.2[a,b]	113.9 ± 9.4[a,b]	1.27 ± 0.2	26.7 ± 2.3[a,b]
2.64 ± 0.1[a,b]	100.0 ± 8.3[a,b]	1.01 ± 0.2[a]	28.2 ± 2.5[a,b]
1.68 ± 0.1[a,b]	100.4 ± 6.4[a,b]	1.01 ± 0.1[a]	18.3 ± 2.2

during the first three stages of the expedition. There was also a significant (62%) increment in the A/NA ratio, suggesting an increased activation of the adrenal medulla by the sympathetic system. The blood serotonin increased considerably over the first two stages of the traverse, but decreased to a subnormal level by stage 4 (table 10.1).

The urinary excretion of adrenaline and NA increased over the expedition (table 10.2). DOPA and DA (the metabolic precursor of NA) and the breakdown products of adrenaline and NA (MN and NMN, free and bound, VMA and HVA) are also increased significantly, the maximal excretion of both catecholamines and their precursors occurring at stage 4 of the expedition (table 10.2, 10.3). In keeping with the blood data, the adrenaline/NA (A/NA) ratio also increased by 52% at stage 1 of the expedition (table 10.4), pointing to the development of an emotional response via the sympathetic nervous system.

Adrenaline methylation, as shown by the MN/adrenaline (MN/A) ratio, and adrenaline binding both decreased over the trek, contributing to the increased effective blood and urinary levels of adrenaline. However, perhaps because the subjects were relatively fit, the stress reaction was apparently quite moderate, as shown by the absence of any consistent changes in NA synthesis (the NA/DA ratio) or of catabolic methylation (the NMN/NA ratio). NA binding to acid radicals also showed little change. The observed increase in circulating levels of NA thus seems to reflect an unchanged metabolism, but an increased release from depots.

DA synthesis (the ratio DA/DOPA) decreased over the trek, while its inactivation (the ratio HVA/DA) increased, especially at stage 4 and after completion of the mission. These various changes, like the increasing deamination of catecholamines, are apparently designed to protect the organism from the increasing amounts of DA that are being mobilized (table 10.4).

We may conclude that the general secretory activity of the sympatho-adrenal system gradually increased over the expedition, and after completion of the trek it still exceeded initial values. The reactivity of the sympathetic system, as gauged by the ratio of the summed catecholamines to that of their free forms, changed in similar fashion (table 10.4). However, increased mobilization of catecholamines was not accompanied by changes in synthesis or degradation, with the exception of dopamine.

The urinary excretion of HM, serotonin, HD and 5-HIAA all increased over the expedition, while trytophan and 5-HOT excretion decreased. The largest changes in the serotoninergic and histaminergic systems were seen at stage 4 of the expedition (table 10.3). The serotonin/HM (S/HM) ratio increased significantly at stage 4 and after completion of the

10 Status of Selected Hormones and Biologically Active Compounds

Table 10.4. Relative metabolic activity indices for urinary catecholamines, their metabolites, serotonin and histidine

Variable	Moscow	Dikson	Passage, days		Ottawa
			12	74	
A/NA	100	123	152	149	119
NA/DA	100	94.5	105	110	113
DA/DOPA	100	99	87.5	85.6	95
MN/A	100	82	56	69	66
NMN/NA	100	105	96	92	86.5
HVA/DA	100	99	124	167	164
VMA/A + NA	100	107	102	136	151
VMA/MN + NMN	100	110	127	155	192
ΣCA, total secretory activity µg/24 h	7,195	8,295	10,886	16,305	14,679
ΣCA/ΣCAf	100	104	118	146	161
Binding of A, %	100	86	71	69	84
Binding of NA, %	100	100.9	100	102	102
Binding of DA, %	100	97	96	95	95
Binding of MN, %	100	98	99.7	89	102
Binding of NMN, %	100	104	104	103	104
S/5 HOT	100	175	264	580	369
5-HOT/T	100	93	113	102	102
5-HIAA/S	100	86	74	45	45.6
HM/HD	100	145	197	217	128
S/HM	100	104	105	150	169

All data are expressed as percent of the initial values. A = Adrenaline; NA = noradrenaline; DA = dopamine; MN = metanephrine; NMN = normetanephrine; HVA = homovanillic acid; VMA = vanillylmandelic acid; ΣCA = total output of catecholamines over 24 h period; ΣCAf = total output of free catecholamines summed over 24 h period; S = serotonin; 5-HOT = 5-hydroxytryptophan; T = tryptophan; 5-HIAA 5-hydroxy-indoleacetic acid; HM = histamine; HD = histidine.

trek, coincident with the changes in sympathetic activity. Serotonin synthesis (the ratio serotonin/5-HOT) increased progressively, reaching a maximum at stage 4 of the trek, while serotonin inactivation (the ratio 5-HIAA/serotonin) decreased gradually, the lowest levels being seen at stage 4 and in Ottawa after completion of the mission. HM synthesis (the ratio HM/HD) increased to a maximum at stage 4.

The data obtained on the present expedition suggest that serotoninergic activity was increased during the first half of the trek. This may reflect an effect of this system upon the excitability of the vasomotor and thermoregulatory centers, as a response to the combination of extreme cold and intensive physical effort.

Conclusions

These various mechanisms and counterreactions are associated with increased blood levels of both NA and serotonin [Vasilyev and Chugunov, 1985]. The approach of mobilizing NA reserves without changing synthesis or metabolism, as seen in the well-trained members of the present expedition, is judged by us as physiologically the most appropriate response to a stressful situation.

Detailed analysis of sympathetic/hormonal links suggests that the participants in the expedition were not overly stressed. The transitory, unstable stage of adjustment [Barbarash and Dvurechenskaya, 1986], seen at the time of the first parachute drop of supplies, decreased as the expedition continued. We suggest that overall acclimatization was attributable to such factors as hyperfunction of the sympatho-adrenal nervous system, activation of nucleic acid and protein synthesis, hypertrophy of brown adipose tissue, and the development of improved transport mechanisms for delivery of oxygen and the oxidation of substrate [Radomski and Boutelier, 1982; Barabarash and Dvurechenskaya, 1986]. Other factors contributing to a satisfactory progression from stage 2 (rapid adjustment) to stable, long-term (stage 3) acclimatization to the overall environment included a high level of physical fitness, the wearing of appropriate clothing, maintenance of a high level of physical activity, an appropriate choice of foods, and other social adjustments.

Acclimatization to cold is associated with other changes in mechanisms for activation of the adrenergic system. In particular, there are changes in the sensitivity of the cold receptors [Hensel and Schäfer, 1982], with a decrease in the maximal activity of brain stem neurones in the cold [Hinckel and Schröder-Rosenstock, 1982], and an increased power and flexibility of central mechanisms of thermoregulation linked to the development of conditioned reflexes [Cooper, 1976; Van Someren et al., 1982]. In the early stages of cold exposure, central regulatory mechanisms induce a stable activation of the adrenergic system, which in turn brings about metabolic and structural changes that enhance heat production. Catecholamines play a key role in the critical uncoupling of respiration from phosphorylation, with a rapid increase in heat production [Skulachev, 1982].

Under the specific conditions of the ski trek, cold exposure was coupled with a high level of physical activity. It appears that the combination of physical and environmental stress contributed to the harmonious development of acclimatization, with increased resistance to injurious responses.

Pituitary/Adrenocortical Axis

General Considerations

The adrenal cortex secretes three main groups of hormones, glucocorticoids (particularly cortisol), mineralocorticoids (particularly aldosterone), and the sex hormones (their contribution being more important in women than in men). Cholesterol is a precursor of progesterone, which in turn can be converted to cortisol or corticosterone. The latter can be further metabolized to aldosterone.

The normal trigger to release of cortisol is ACTH. Heavy, prolonged exercise may increase both blood and urinary cortisol levels, particularly if the activity is perceived as stressful. Some 95% of cortisol is transported in the blood in bound form, so that activity can be substantially modified by changes in secretion or excretion. This hormone penetrates key cells, enhancing the activity of rate-limiting enzymes. In particular, it mobilizes both peripheral fat depots and protein stores, thus conserving carbohydrate.

Changes over the Course of the Expedition

Blood ACTH levels of the participants decreased (table 10.5), especially by the second half of the expedition (days 55 and 74). Perhaps because of initial anxieties, cortisol readings in Moscow were very high. Blood cortisol increased relative to the more moderate Dikson baseline over the first half of the trek, but subsequently dropped back to these initial values.

The urinary excretion of cortisol was significantly increased up to day 55 of the expedition, but by day 74 it had returned to initial values. Concentrations of precursor progesterone were increased, especially at days 12 and 74. The ratio of ACTH to cortisol was lower than the initial Dikson values on days 12, 29 and 74 (table 10.5–10.7).

The first two stages of the expedition were thus marked by an increase in the glucocorticoid function of the adrenals, as shown by the increased blood cortisol levels, despite the unchanged ACTH levels. The unchanged ACTH may reflect either negative feedback from the rising cortisol levels or an altered sensitivity of the hypothalamus/hypophysis to such feedback. In either event, the increased concentration of the 'working' hormone (corti-

Table 10.5. Blood concentrations of some pituitary and adrenal gland hormones as well as somatostatin and somatomedin (mean ± SD)

Location	ACTH pg/ml	Cortisol nmol/l	Growth hormone ng/ml
Moscow	38.9 ± 4.8	530.2 ± 32.9	0.62 ± 0.09
Dikson	45.9 ± 5.5	381.6 ± 32.6[a,b]	0.55 ± 0.05
Day 12	33.8 ± 2.0	433.9 ± 17.5[b]	4.59 ± 1.13[a,b]
Day 29	39.2 ± 5.8	480.2 ± 23.4	2.19 ± 0.64[a,b]
Day 55	31.1 ± 2.3[b]	310.9 ± 31.7[a]	3.29 ± 0.67[a,b]
Day 74	26.7 ± 1.9[a,b]	351.1 ± 29.5[a]	0.74 ± 0.16
Ottawa	32.3 ± 3.1	318.8 ± 38.3[a]	1.73 ± 0.38[a,b]

[a,b]Significant differences relative to baseline tests conducted in Moscow (a) and data collected in Dikson (b).

Table 10.6. Urinary excretion of steroid hormones, cyclic nucleotides and electrolytes with urine osmolality (mean ± SD)

Location	Cortisol µg/24 h	Aldosterone µg/24 h	cAMP µmol/24 h
Moscow	50.2 ± 6.3	5.76 ± 0.54	3.39 ± 0.44
Dikson	81.9 ± 8.8[a]	8.21 ± 0.56[a]	5.82 ± 0.31[a]
Day 12	124.7 ± 21.6[a]	8.49 ± 1.03[a]	3.85 ± 0.36[b]
Day 55	168.0 ± 22.8[a,b]	15.4 ± 1.12[a,b]	5.74 ± 0.37[a]
Day 74	70.9 ± 14.0	8.25 ± 0.89[a]	5.90 ± 0.52[a]

[a,b]Significant differences relative to baseline tests conducted in Moscow (a) and data collected in Dikson (b).

Table 10.7. Key ratios for blood levels of hormonal and biologically active compounds (mean)

Location	ACTH: β-endorphin	Cortisol: IRI	TSH: T_3	TSH: fT_3	fT_3: rT_3	fT_3: T_3
Moscow	6.71	1.78	0.78	0.23	0.44	3.78
Dikson	5.57	1.00	1.01	0.25	0.35	3.83
Day 12	5.68	2.54	1.10	0.32	0.35	3.01
Day 29	7.64	1.98	1.84	0.47	0.32	3.01
Day 55	7.24	2.52	0.73	0.30	0.37	2.88
Day 74	9.20	0.91	0.93	0.32	0.34	2.87
Ottawa	10.00	1.80	0.68	0.26	0.44	3.08

IRI = Immunoreactive insulin

Progesterone nmol/l	β-Endorphin pmol/l	Somatostatin pg/ml	Somatomedin nmol/l
0.63 ± 0.03	5.49 ± 0.4	4.49 ± 0.6	25.0 ± 1.8
0.70 ± 0.08	6.79 ± 1.1	3.61 ± 0.2	24.5 ± 1.1
0.92 ± 0.11[a]	7.46 ± 0.9	4.56 ± 0.4	19.9 ± 1.3[a,b]
0.55 ± 0.05	3.65 ± 0.5[a,b]	5.99 ± 0.6[b]	18.2 ± 1.1[a,b]
0.60 ± 0.08	4.53 ± 0.6	4.05 ± 0.2	20.7 ± 1.5
1.51 ± 0.07[a,b]	2.22 ± 0.3[a,b]	4.32 ± 0.4	24.8 ± 1.3
0.85 ± 0.09[a,b]	3.39 ± 0.4[a,b]	4.38 ± 0.3[b]	20.1 ± 1.4[a,b]

cGMP μmol/24 h	Sodium mmol/24 h	Potassium mmol/24 h	Osmolality mosm/kg
1.87 ± 0.25	181.4 ± 20.9	66.7 ± 3.2	899.5 ± 46.9
2.46 ± 0.22	383.0 ± 29.3[a]	85.6 ± 6.4[a]	867.5 ± 24.5
1.30 ± 0.16[b]	172.1 ± 20.9[b]	78.3 ± 8.3	786.5 ± 70.7
1.79 ± 0.17[b]	308.1 ± 29.7[a]	93.5 ± 7.5[a]	885.0 ± 42.1
1.47 ± 0.22[b]	357.4 ± 26.0[a]	61.7 ± 8.0[b]	809.2 ± 34.6

fT_4 : T_4	T_3 : rT_3	T_4 : T_3	cAMP: cGMP	ALD: renin	ACTH: cortisol	ACTH: ALD	C-peptide: IRI
0.16	0.12	57	4.68	165	0.08	0.34	0.04
0.15	0.09	58	2.70	130	0.15	0.38	0.04
0.06	0.12	127	29.4	121	0.07	0.19	0.05
0.08	0.11	97	33.0	97	0.07	0.30	0.05
0.12	0.13	53	21.2	90	0.10	0.35	0.06
0.13	0.22	57	9.16	45	0.09	0.21	0.03
0.15	0.14	43	2.05	273	0.12	0.40	0.06

sol) with an unchanged concentration of the central regulating hormone (ACTH) indicates a change in homeostatic mechanisms in response to the conditions encountered on the trek. At this stage, there were no changes in blood levels of the precursor progesterone.

During the second half of the expedition, the intensification of adrenocortical activity gave way to a phase of inhibition. Blood levels of cortisol decreased. This apparently reflected a combination of decreased production and intensive catabolism and/or excretion of steroids. Urinary excretion of cortisol was high at stage 3, but had decreased at stage 4, suggesting that at this stage secretion and excretion were once again equilibrated. ACTH levels were below initial readings during the second half of the traverse (table 10.5). The ratio of ACTH to cortisol thus increased (table 10.7), although remaining below initial levels, showing that the sensitivity of the adrenal cortex to central stimulation was still greater than initially. This finding seems evidence of chronic stress [Dilman et al., 1987]. The problem seems not an exhaustion of the adrenal cortex, but rather a deficiency of adrenocortical drive. This drive is derived from the limbic system (particularly the hippocampus), and is perhaps reduced in an attempt to avoid adrenal exhaustion during prolonged physical activity [Viru, 1975; Viru et al., 1987].

Blood levels of progesterone had increased sharply by stage 4 of the trek, but cortisol production was not increased at this stage. The build-up of the precursor suggests that a defect in hydroxylation may have developed at the level of carbon atoms 11 or 17. Similar changes, including a reduction in the ratio of urinary 17-hydroxyketosteroids to 17-dehydroxyketosteroids, have been described previously during prolonged muscular work [Bugard et al., 1958]. According to this earlier research, a hypersecretion from the adrenal cortex has a negative influence upon hydroxylation at the 17 carbon atom.

After completion of the expedition, indices characterizing the function of the hypothalamus/hypophysis/adrenocorticoid axis had all returned to initial levels.

Renin/Angiotensin/Aldosterone Axis

General Considerations

The usual sequence of events during vigorous exercise is a decrease in renal blood flow, triggering the release of renin and thus angiotensin. This in turn stimulates aldosterone release from the adrenal cortex. The aldosterone promotes sodium retention and potassium elimination. Other possible triggers to aldosterone release are ACTH, other hormones from the pituitary, and changes in sodium and potassium ion concentrations. Vasopressin is a further pituitary hormone that potentially influences the circulatory adaptation to heavy exercise; in physiological doses, it acts mainly as a potent antidiuretic.

Changes Observed during the Expedition

As predicted above, the renin/angiotensin/aldosterone axis was generally stimulated during the trek (table 10.6–10.8). Plasma renin activity was increased at stage 1, and especially at stage 4 of the expedition when it reached more than three times the initial level. Blood aldosterone concentrations were also increased at stage 1, but not at other points during the traverse. Blood levels of angiotensin-II (the active form of angiotensin) increased progressively over the course of the trek, and by stage 4 had reached 6.4 times the original values.

Vasopressin level also increased very steeply, particularly during the first half of the expedition, values being 4.4 times the initial level at day 12, and 20 times initial at day 29, but dropping back to 8.8 times background at stage 3, and 2.8 times at stage 4. The activity of the ACE was decreased throughout the expedition.

Plasma sodium and potassium levels were somewhat at variance with what might have been expected from the powerful activation of the aldosterone axis. Osmolality was decreased at stage 1 (table 10.6), this being associated with a small decrease in plasma sodium concentration. Potassium concentrations were substantially increased both during and following the expedition.

The osmolality of the urine (table 10.6) showed a statistically insignificant trend to decrease at the beginning of the expedition. Sodium excretion was decreased at stage 1, and potassium excretion was increased over the first three stages but was decreased at stage 4. Aldosterone excretion was increased, particularly at stage 3. The responsiveness of the adrenal cortex, as shown by the ratio of aldosterone/renin gradually decreased over the course of the trek (table 10.7), especially at stage 4, but there was a sharp increase after completion of the mission (tables 10.6–10.8).

We may conclude that there was a substantial increase in activity of the renin/angiotensin/aldosterone axis at stage 1 of the trek; given the dissonance with plasma minerals, this may possibly represent an attempt to compensate for the decrease in plasma sodium and the increase in plasma potassuim seen at this stage. The decrease in plasma sodium was associated with a decrease in plasma osmolality, and thus an apparent need for diuresis; however, vasopressin levels not only failed to decrease, but even increased quite markedly, suggesting the dominance of nonosmotic stimuli in the regulation of vasopressin secretion at this stage.

Table 10.8. Blood levels of renin, angiotensin, aldosterone and vasopressin with related electrolytes (mean ± SD)

Location	Renin ng/ml/h	Aldosterone pg/ml	Vasopressin pg/ml
Moscow	0.86 ± 0.08	115.4 ± 5.9	–
Dikson	1.05 ± 0.12	113.4 ± 7.1	2.03 ± 0.23
Day 12	1.77 ± 0.24[a,b]	204.5 ± 12.0[a,b]	8.9 ± 2.1[b]
Day 29	1.06 ± 0.15	106.2 ± 11.5	39.9 ± 7.2[b]
Day 55	1.36 ± 0.13[a]	131.9 ± 7.9	17.8 ± 3.7[b]
Day 74	3.49 ± 0.49[a,b]	133.2 ± 12.8	5.7 ± 1.2[b]
Ottawa	0.59 ± 0.11[b]	94.8 ± 11.0	2.08 ± 0.23

ACE = Angiotensin-converting enzyme.
[a,b]Significant differences relative to baseline tests conducted in Moscow (a) and data collected in Dikson (b).

The decrease in urinary sodium excretion in the early part of the trek seems to be explained by the reduction in plasma sodium concentrations, but the decrease in urinary potassium in the face of increased plasma levels of potassium suggests the initiation of some potassium-sparing mechanism in response to the combination of cold stress and vigorous physical activity. Previous observations on a shorter ski trek also demonstrated a decreased excretion of sodium and potassium, with a rise in plasma potassium that was linked to an increased blood level of aldosterone [Krylov and Tigranian, 1986].

Previous studies of hypothermia noted a sodium loss due to cold diuresis [Bass, 1954; Smith and de Long, 1975; Slepushkin and Stepnoi, 1980; Mostovoy, 1982; Seliatitskaia et al., 1985], an increased [Slepushkin and Stepnoi, 1980; Mostovoy, 1982] or unchanged [Bass, 1954; Smith and de Long, 1975] potassium loss, increased blood aldosterone [Gordon et al., 1967] and plasma renin activity [Smith and de Long, 1975; Gordon et al., 1967], and unchanged [Gibbs, 1984] or decreased blood vasopressin [Mostovoy, 1982, Seliatitskaia et al., 1985], with an increased output of ADH from the supraoptic and paraventricular hypothalamic nuclei, and an outflow of vasopressin along the axons, with accumulation in the neurohypophysis (Saakov et al., 1969). However, it is difficult to compare any such data with the present findings, since the earlier exposures were of shorter duration and did not involve comparable physical stress.

Since the osmolality of the blood was low throughout the expedition, the stimulus to the high blood levels of vasopressin must have been

Angiotensin II pg/ml	ACE nmol/ml/min	Sodium mmol/l	Potassium mmol/l
–	–	–	–
12.2 ± 2.4	96.1 ± 3.8	141.3 ± 0.9	4.17 ± 0.05
36.7 ± 7.0[b]	62.0 ± 5.3[b]	137.6 ± 1.1[b]	5.3 ± 0.12[b]
28.5 ± 1.4[b]	71.8 ± 3.2[b]	139.6 ± 0.8	4.65 ± 0.08[b]
55.9 ± 14.3[b]	80.0 ± 3.7[b]	139.4 ± 0.7	4.65 ± 0.09[b]
77.9 ± 7.2[b]	71.7 ± 3.4[b]	140.3 ± 0.7	4.97 ± 0.1[b]
7.8 ± 0.9	87.9 ± 6.2	139.2 ± 0.8	4.82 ± 0.1[b]

nonosmotic. Intensive physical activity is known to increase the secretion of vasopressin [Onaka et al., 1986a, b], probably for reasons other than shifts in water and salt metabolism [Smelik, 1984]. Various factors associated with the expedition (sudden changes of situation, physical discomfort, an unusual diet, sufficient physical strain to assume an emotional coloring, limited interpersonal contacts, unfamiliar colleagues with a language barrier, and the monotony of skiing) may all have contributed to stress, with the requirements of nonspecific and specific adaptive responses to the increased demands on the system [Michajlovskij et al., 1988]. Prolonged exposure to low temperatures may also have had a direct effect on vasopressin secretion. Acute cold exposure usually decreases vasopressin secretion, but it is still conceivable that the very prolonged exposure provoked an increase. In the present instance, the vasopressin secretion should probably be viewed as a stress rather than a cold reaction; the presence of such a stress was certainly indicated by the high catecholamine levels (table 10.1) and the low levels of serum testosterone (table 10.9). A further immediate stimulus may have been a change in blood volume or the development of hypoglycemia [Baylis et al., 1981]. Blood glucose levels were low throughout the expedition and the immediate recovery period (table 10.1). As the expedition continued, there may also have been a downregulation of vasopressin receptors [Roy et al., 1981], the sustained increase in concentrations of the hormone leading to either a decrease in the numbers of vasopressin receptors or of their affinity for the hormone. A decrease in the number or affinity of receptors would also explain the high levels of free vasopressin in the blood stream.

Table 10.9. Blood levels of hormones characterizing the pituitary-gonadal system (mean ± SD)

Location	Prolactin μIU/ml	Luteotropin mIU/ml	FSH mIu/ml	Testosterone nmol/l	Estradiol nmol/l	Estriol nmol/l
Moscow	331.1 ± 53.3	5.69 ± 0.42	2.23 ± 0.39	19.0 ± 1.2	0.14 ± 0.008	0.84 ± 0.03
Dikson	367.5 ± 54.8	8.32 ± 0.48[a]	5.24 ± 0.57[a]	20.1 ± 1.1	0.15 ± 0.006	0.86 ± 0.03
Day 12	50.0 ± 3.9[a,b]	5.78 ± 0.68[b]	4.17 ± 0.78	21.4 ± 2.0	0.14 ± 0.004	0.85 ± 0.03
Day 29	40.7 ± 5.0[a,b]	6.43 ± 0.91	2.62 ± 0.18[b]	13.0 ± 1.4[a,b]	0.13 ± 0.009	0.93 ± 0.08
Day 55	46.4 ± 3.9[a,b]	5.99 ± 0.42[b]	4.22 ± 0.57[a]	13.1 ± 1.2[a,b]	0.14 ± 0.006	0.31 ± 0.01[a,b]
Day 74	44.3 ± 1.9[a,b]	5.55 ± 0.41[b]	4.13 ± 0.58[a]	14.6 ± 1.0[a,b]	0.17 ± 0.009[a]	0.40 ± 0.02[a,b]
Ottawa	80.2 ± 7.3[a,b]	6.47 ± 0.18	4.15 ± 0.39[a]	19.9 ± 0.8	0.16 ± 0.006	0.34 ± 0.01[a,b]

[a,b]Significant differences relative to baseline tests conducted in Moscow (a) and data collected in Dikson (b).

Vasopressin secretion is inhibited by β-endorphin (BE) [Knepel et al., 1982, 1985] and cathecholamines [Armstrong et al., 1982; Everett et al., 1983]. Thus a full analysis must take account of the declining concentration of BE (table 10.5) and the pronounced increase in catecholamines during and following the expedition (table 10.1). Possibly, these inhibitory mechanisms are modified by the combination of prolonged cold and sustained exercise.

The physiological 'purpose' of the vasopressin release is unclear. Possibly, it reflects the intense stressing of the cardiovascular system by the demands of the expedition . The output seemed greatest during that part of the expedition when cold was greatest. It may have served as a nonspecific response to a substantial water loss [Vigas et al., 1980]. Irrespective of functional purpose, the amount of secretion suggests a high level of stress [Bunag et al., 1967].

As the expedition continued, blood levels of potassium and sodium stabilized. This may imply that adjustment of water and salt metabolism was now relatively complete, the balanced state being marked by some increase in plasma potassium concentrations. Nevertheless, some changes in the regulating systems were observed at stage 3 of the trek, including a slight increase in renin activity, a significant increase in angiotensin II, and a large increase in the urinary excretion of aldosterone (table 10.6). These changes probably contributed to the normalization of plasma sodium and potassium levels. By stage 4 of the expedition, the hours of daylight increased, there was a substantial increase in ambient temperature, and the physical strain imposed by the environment became less intense. At stage 4, the decrease in blood vasopressin concentration would have contributed to the continuing increase in renin and angiotensin-II concentrations [Bunag et al., 1967; Vander, 1968; Vandongen, 1975; Khokhar et al., 1976; Hesse and Nielsen, 1977; Morton et al., 1982], probably through an increase in blood volumes [Smith et al., 1979a].

The aldosterone/renin coefficient decreased successively over the expedition, suggesting a decreased role of angiotensin-II in the regulation of aldosterone secretion (table 10.7). It may be concluded that initially the renin/angiotensin/aldosterone axis was controlled by ionic metabolism and served to stabilize ionic concentrations. However, as the strains of the expedition were encountered, the activity of this axis became regulated by other factors, including the decreasing levels of vasopressin and/or responses of the cardiovascular system to the new conditions.

The activity of the renin/angiotensin/aldosterone axis decreased sharply after completion of the mission, despite normal plasma levels of sodium and potassium. This suggests that there may have been a decrease in sympathetic drive to renin secretion at this stage in the mission [Reid et al., 1978].

ACE activity was apparently unrelated to changes in either plasma renin activity or blood concentrations of angiotensin-II, supporting the view that there are independent control mechanisms for renin secretion and ACE activity [Horky et al., 1971]. The decrease in ACE activity at stage I of the trek can probably be linked to the known effect of a sodium deficit upon blood levels of converting enzyme [Kotchen and Roy, 1987].

Hypothalamus/Hypophysis/Gonadal Axis

General Considerations

In the context of exercise, prolactin mobilizes fat and reduces the urine output. LH is a pituitary hormone that in the male promotes testosterone secretion by the interstitial cells of the testis. Testosterone has an important role in counteracting the catabolic action of cortisol during prolonged exercise; typically, stress reduces testosterone levels [Repcekova and Mikulaj, 1977; Gray et al., 1978; Tigranian et al., 1980]. Testosterone is broken down to estradiol and then to estriol during excretion.

Changes over the Expedition

Blood prolactin levels decreased during and following the expedition (table 10.9). Blood concentrations of LH were also reduced except at stage 2 (day 29), whereas FSH was increased only by day 29 of the expedition. Testosterone levels were reduced on days 29, 55 and 74, whereas blood concentrations of estradiol showed no changes. Levels of estriol (an estradiol metabolite) were unchanged in the first half of the expedition, but declined in the second half (table 10.9). During the second half of the expedition, there was thus a substantial decrease in testicular activity, providing one more indication of cumulative stress.

It is less certain whether the deficit of testosterone production arose through a suppression of function in the hypothalamus/hypophysis/gonadal axis, or a more direct action upon the steroid-synthesizing gland. However, the authors suspect a decrease in sensitivity of the testes to LH [Katsia et al., 1984], probably related to a decrease in local blood flow.

During the recovery period, gonadal function was restored to initial levels, although the trophic action of LH was still decreased. Low blood levels of estradiol suggested that the metabolism of testosterone also remained disturbed.

Prolactin secretion was decreased throughout the expedition. The standard reaction to stress is an increased secretion of prolactin [Mills and Robertshaw, 1983] due to a decrease in hypothalamic DA (which has an inhibitory effect upon prolactin secretion) [Kamberi et al., 1971]. The decrease in DA, NA and serotonin also lead to an increased secretion of ACTH [Scapagnini and Preziosi, 1973]. The fact that the secretion of both ACTH and prolactin decreased during the present expedition suggests that hypothalamic levels of DA, NA and serotonin may have increased. Various authors have also noted a decreased prolactin level with cold exposure [Mills and Robertshaw, 1981; Kalita et al., 1984; O'Malley et al., 1984; Tigranian et al., 1985, 1988a, b; Krylov and Tigranian 1986; Leppäluoto et al., 1988], suggesting that the response to extreme cold may differ from reactions to other forms of stress.

In most mammals, an increase in prolactin levels is associated with decreased concentrations of gonadotropins [McNeilly, 1987]. However, in the present study there was a parallel decrease in prolactin and LH, again suggesting that the cold stress had disturbed the normal interrelationship between these two hormones.

Hypothalamus/Hypophysis/Thyroid Axis

General Considerations

Thyroid hormone (thyroxine T_4) could make a contribution to cold acclimatization by uncoupling the link between carbohydrate metabolism and adenosine triphosphate formation, stimulating the level of basal metabolism [Shephard, 1983]. It also increases the lipolytic effect of NA, it can cause an accumulation of cAMP within adipose cells in its own right, and it promotes cardiac hypertrophy. Exercise is known to increase the secretion of thyroxine and also to reduce protein binding, while training enhances both the secretion and the degradation of T_4. Very prolonged cold exposure also increases T_4 levels [Shephard, 1983].

In the present study, measurements were made of the pituitary thyroid-stimulating hormone (thyrotropin, TSH), total T_4, fT_4, the more rapidly acting T_3, fT_3, and the inactive form of T_4, rT_3. TSH stimulates thyroid function, while the T_4 has a sluggish negative feedback upon the hypophyseal production of TSH.

Changes Observed during the Expedition

The blood concentrations of TSH were only increased significantly on day 29 of the expedition (table 10.10). Blood concentrations of T_3 did not change, but fT_3 decreased at all stages of the trek relative to initial values in Moscow and Dikson. Blood concentrations of rT_3 did not change, but total T_4 increased in the first half of the journey, returning to baseline values for the second half of the trek. fT_4 decreased at stages 1–3 of the traverse.

Hormone Ratios

The ratio TSH/T_3 behaved in a similar manner to TSH itself, with a sharp increase at stage 2 of the expedition (table 10.7, 10.10). During stage 1, the ratio T_4/T_3 increased; total T_4 increased, but total T_3 was unchanged, suggesting that there was a decreased formation of T_3 from T_4, due to the excess of glucocorticoids at this stage of the trek [Burr et al., 1976] (table 10.5). fT_4 decreased at this stage of the expedition; fT_3 also decreased despite unchanged levels of rT_3 and T_3 (possibly indicating a decreased function of the thyroid gland) [Shcherbakova and Rom-Boguslavskaya, 1988]. An increase in the ratio T_3/rT_3 (table 10.7) suggests an attempt to sustain an euthyroid state despite the drop in fT_3 and fT_4 [Goltseva et al., 1987; Fabri and Pashchenko, 1987]. The ratio fT_4/T_4 showed a pronounced decrease, despite the increase in T_4. TSH was unchanged, and the inactive form of T_4 (rT_3) was unchanged. Given the unchanged TSH and decreased levels of free forms of the thyroid hormone, any increase in total T_4 seems due to increased binding by thyroglobulins and a decreased clearance of thyroid hormones from the plasma, rather than an increased secretion.

The changes observed at stage 2 of the trek were similar to stage 1, with the exception of an increase in TSH. The ratio of TSH/T_3 thus increased noticeably during this segment of the traverse. The increased secretion of TSH may reflect a combined response to the cortisol-induced inhibition of the hypophyseal reaction to thyrotropin-releasing factor [Sowers et al., 1977a, b] a decreased feedback from T_4, and an increased feedback from adrenaline, serotonin (table 10.1) and cAMP (table 10.11). The secretion of TSH is stimulated by adrenaline and serotonin [Abrahamson et al., 1987] and cAMP [Davis and Sheppard, 1987]. The increased

Table 10.10. Blood levels of hormones characterizing the pituitary-thyroid axis (mean ± SD).

Location	TSH mIU/l	T_3 nmol/l	Free T_3 pmol/l	Reverse T_3 ng/100 ml	T_4 nmol/l	Free T_4 pmol/l
Moscow	1.20 ± 0.13	1.58 ± 0.06	5.97 ± 0.32	13.4 ± 0.6	90.3 ± 2.6	14.0 ± 0.52
Dikson	1.38 ± 0.06	1.56 ± 0.07	5.98 ± 0.23	16.9 ± 1.8	91.2 ± 3.2	13.3 ± 0.65
ay 12	1.59 ± 0.25	1.56 ± 0.08	4.70 ± 0.18[a,b]	13.5 ± 1.0	197.6 ± 37.4[a,b]	11.7 ± 0.43[a,b]
Day 29	2.10 ± 0.33[a,b]	1.43 ± 0.08	4.31 ± 0.13[a,b]	13.4 ± 1.2	139.1 ± 15.5[a,b]	10.5 ± 0.76[a,b]
Day 55	1.37 ± 0.11	1.73 ± 0.10	4.98 ± 0.24[a,b]	13.4 ± 1.0	91.2 ± 3.8	10.7 ± 0.88[a,b]
Day 74	1.77 ± 0.19[a]	1.65 ± 0.11	4.73 ± 0.18[a,b]	13.9 ± 1.2	93.8 ± 7.1	12.0 ± 0.79
Ottawa	1.39 ± 0.12	1.77 ± 0.08	5.46 ± 0.22	12.4 ± 0.8	75.8 ± 2.1[a,b]	11.7 ± 0.45[a]

[a,b]Significant differences relative to baseline tests conducted in Moscow (a) and data collected in Dikson (b).

Table 10.11. Blood levels of hormones regulating carbohydrate and calcium metabolism and blood osmolality (means ± SD).

Location	Insulin pmol/l	C-peptide ng/ml	Glucagon pg/ml	Parthyroid hormone pg/ml	Calcitonin pg/ml	Osmolality mosm/kg
Moscow	40.4 ± 10.6	1.51 ± 0.09	44.0 ± 7.3	62.7 ± 4.3	–	–
Dikson	37.1 ± 4.0	1.67 ± 0.08	33.1 ± 4.9	48.6 ± 3.7[a]	39.8 ± 1.7	277.5 ± 1.7
Day 12	39.0 ± 4.2	1.72 ± 0.09	45.0 ± 5.2	71.5 ± 7.3[b]	38.4 ± 1.7	267.1 ± 1.5[b]
Day 29	54.0 ± 4.0[b]	1.82 ± 0.06[a]	56.3 ± 7.2[a]	81.4 ± 7.0[a,b]	41.0 ± 1.6	274.7 ± 1.8
Day 55	22.2 ± 2.7[b]	1.25 ± 0.07[a,b]	29.5 ± 5.2	75.4 ± 4.6[b]	41.4 ± 2.4	274.5 ± 1.0
Day 74	49.1 ± 4.0	2.17 ± 0.11[a,b]	74.4 ± 7.2[a,b]	68.9 ± 6.4[b]	41.3 ± 1.7	277.9 ± 1.1
Ottawa	23.4 ± 1.6[b]	1.26 ± 0.03[a,b]	34.7 ± 3.7	25.0 ± 2.3[a,b]	34.0 ± 1.4[b]	272.6 ± 1.4

[a,b]Significant differences relative to baseline tests conducted in Moscow (a) and data collected in Dikson (b).

TSH/T_3 ratio suggests that the response threshold of the hypothalamus/hypophyseal/thyroid axis has been increased [Dilman et al., 1987]. Considering the ability of TSH to activate not only the synthesis and secretion of the T_4 in the thyroid gland, but also peripheral deiodination [Ikeda et al., 1986; Watson-Whitmyre and Stetson 1983], the change in deiodination at this stage (the decrease in T_4/T_3) may be due simply to the increased blood level of TSH. The T_4/T_3 ratio nevertheless still exceeded initial values, suggesting a continuing relative T_3 deficiency. The level of T_4 still exceeded initial values, but was lower than at stage 1 of the trek.

By stage 3 of the traverse, TSH levels had returned to their initial values, but the blood content of fT_3 and T_4 remained decreased. A decrease in the ratio of TSH/T_3 developed (table 10.7), which may indicate a lowering of the response threshold for the hypothalamus hypophysis complex. Total T_4 decreased, while total T_3 increased; the ratio T_4/T_3 thus returned to initial levels, and there was no longer a relative T_3 deficiency. These changes probably indicate an accelerated peripheral conversion of T_4 to T_3. If so, most of the T_4 was converted to the inactive form (rT_3) rather than the active form of T_3. The ratio fT_3/rT_3 decreased, despite constant levels of rT_3 in peripheral blood. The main feedback link in the thyroid axis is provided by a change in the sensitivity of the hypophyseal cells to thyrotropin-releasing hormone; this depends mainly on the T_3 concentration [Mirakhmedov, 1979; Starkova, 1981]. A slight decrease in TSH secretion may thus be a natural response to increases in free and total T_3 at this stage of the trek.

During stage 4 of the expedition, none of the thyroid indices differed from their initial values, with the exception of a decrease in fT_3 (table 10.7).

Following the trek, levels of rT_3, free and total T_4 were all decreased relative to the values before departure, total T_4 being lower, and fT_3 higher than at stage 4 of the traverse. Such changes suggest an intensification of thyroid secretory activity, linked to the increase in total T_3 during the period of readaptation to normal climates. This view is confirmed by a noticeable decrease in the compensation index (the ratio of total T_4 to total T_3), despite values for T_3 and T_4 that both lie within the normal range. There may be a peripheral conversion of T_4 to the more active T_3, this process being facilitated by a decrease in inhibitory feedback from rT_3. The continuing low levels of fT_4 indicate an alteration in thyroglobulin binding, with the implication of diminished metabolism of the hormone in the liver and other tissues.

Conclusions

The thyrotropic function of the hypophysis expedition. At all other stages, including the recovery period, the function of the hypophysis remained at the level seen before the expedition. Overall, there was some early inhibition of thyroid function, the most distinctive feature being an active inhibition of deiodination to T_3.

Somatotropin, Somatostatin and Somatomedin-C

General Considerations

Somatotropin, or GH, is secreted by the hypophysis. Release is controlled by somatostatin (inhibitory) and somatomedin (stimulatory) [Alekseev, 1977; Vasilyeva, 1981; Ross et al., 1987]. Somatotropin stimulates the production of somatomedin-C in the liver, and the latter substance stimulates the sulfation of chondroitin and thus the production of bone and cartilage. Somatostatin is also produced by the hypothalamus; it inhibits not only the secretion of growth hormone, but also that of insulin and glucagon.

Growth hormone makes an essential contribution to prolonged exercise by mobilizing necessary lipids. By intensifying lipolysis, it provides an alternative fuel and increases the glucose content of the blood. It may also contribute to the adaptive hypertrophy of muscle seen with sustained activity [Viru, 1981]. Others have previously described increased somatotropic activity during prolonged northern ski expeditions [Kalita et al., 1984, 1987; Krylov and Tigranian, 1984, 1986; Krylov et al. 1985, Tigranian et al., 1985, 1988a, b]

Changes Observed during the Expedition

The growth hormone content of the blood increased sharply over the first three stages of the expedition (table 10.5). By day 74, concentrations had reverted to the initial level, but final readings were again high. The negative feedback from somatostatin increased by day 29, and was also high after completion of the mission, while somatomedin values were low at days 12, 29 and the final sampling (table 10.5).

The intense muscular activity needed in skiing seems a powerful stimulant of GH secretion [Vigas, 1984, 1985]. The present findings certainly show that the expedition brought about a sharp activation of somatotropic function, and that the secretion of GH reverted to the initial levels only at the final stage of the traverse. The GH-induced lipolysis increases blood levels of glycerol and free fatty acids, but there is a simultaneous increase in triglyceride synthesis, using triose phosphates formed from glucose via the glycolytic cycle; this contributes to low blood levels of glucose. Blood glucose was low throughout the expedition (table 10.1), although this reflected mainly high insulin levels (table 10.11) rather than feedback from the GH axis.

The hypoglycemia observed throughout the expedition stimulates GH secretion. However, there was a decrease in blood somatomedin and an increase in somatostatin during the first half of the expedition, with a paradoxical sharp increase in GH output. It would seem that normal mechanisms controlling GH secretion were not operative under the extreme conditions of the transpolar ski trek.

At stage 2 of the traverse, there was a small decrease in GH levels; in this case, with the anticipated accompanying increase in somatostatin and a decrease in somatomedin. This was viewed as evidence that some possible adjustment had occurred to the continuing strain of the journey.

At stage 3, GH levels were again increased; this may be related to the fact that participants were complaining of hunger by this point in the traverse.

After completion of the expedition, the blood GH levels were still increased against a background of increased somatostatin and decreased somatomedin. At this stage, both insulin and glucose levels were decreased.

Interactions with Other Hormones

In terms of potential interactions with other hormones, thyroid hypofunction could decrease the GH secretion [Eastman and Lazarus, 1973; Martynenko et al., 1974]; however, there was no strong evidence of hypothyroidism in the present experiments. There is no consensus as to the action of glucocorticoids on GH secretion. Any feedback probably occurs by inhibiting the production of somatomedin [Pecile and Müller, 1966]. Certainly, at stages 1 and 2 of the expedition, there was evidence that high cortisol levels were linked to a decreased blood level of somatomedin.

Parathormone/Calcitonin System

General Considerations

Feedback mechanisms link the activity of the parathyroid glands and the C cells of the thyroid to blood calcium levels [Drzhevetskaya and Drzhevetsky, 1983]. However, the normal feedback mechanisms may be distorted in some physiological and pathological states (Tsibizov, 1979; Zoloev et al., 1987). One of the regulators of both parathyroid and C cells may be the adrenergic system [Vora et al., 1978], leading to increased outputs of both parathyroid hormone and calcitonin under conditions of stress. The concentrations of the calcium-regulating hormones are known to be increased by a variety of factors, including physical effort, immobilization, emotional stress, and hypoglycemia [Mishina, 1981; Grigoriev et al., 1982; Zoloev, 1983; Drzhevetskaya and Drzhevetsky, 1981]; severe trauma and myocardial infarction both decrease parathyroid hormone levels, while the calcitonin levels increase or remain unchanged [De Boer et al., 1981; Zoloev et al. 1983]. Adrenergic activity increases parathyroid hormone and decreases calcitonin levels [Mayer et al., 1979; Queener et al., 1980; Zoloev et al., 1987]. Further it is known that calcitonin can inhibit catecholamine release [Dupuy et al., 1977].

Changes over the Course of the Expedition

The present experiments featured a substantial release in catecholamines during the trek, against the background of a seemingly invariant level of calcitonin (table 10.11). Following the mission, calcitonin concentrations decreased. Plainly, the feedback mechanism linking catecholamines and calcitonin is more complex than some have suggested; possibly, it also involves levels of calcium ions.

Blood levels of parathormone increased during the expedition, suggesting an increase in parathyroid activity. Subsequent to the mission, parathyroid hormone levels dropped below initial values.

Insulin/C-Peptide/Glucagon System

General Considerations
The synthesis of insulin from the a and b chains of amino acids produced in the β cells of the islets of Langerhans depends upon the temporary insertion of a connecting polypeptide chain (C-peptide). Insulin promotes the entry of glucose, some other sugars and amino acids into muscle and fat cells. It also inhibits the breakdown of glycogen and fat, and decreases the rate of gluconeogenesis.

Glucagon is secreted by the α cells of the pancreatic islets. It has the opposite effect to insulin, in particular activating hepatic enzymes that cause glycolysis and gluconeogenesis.

Changes during the Expedition
Relative to the Dikson baseline blood levels of insulin, C-peptide, glucagon and cortisol were all increased in the first two stages of the trek (table 10.11), with a subsequent decrease in insulin, glucagon and C-peptide in stage 3, and the return of cortisol to its initial value. Most probably, the intense and prolonged muscular work stimulated intramuscular oxidation of free fatty acids and an insulin-dependent consumption of glucose [Nikolsky and Troshin, 1973].

Blood levels of insulin, C-peptide and glucagon were again elevated at stage 4. However, at stage 3 and following completion of the mission, insulin and C-peptide levels were below the values before the expedition. The ratio of the C-peptide to insulin (which characterizes the metabolic clearance of insulin by the liver) [Kosovskii et al., 1988] was thus increased at day 55 and after completion of the mission (table 10.7, 10.11). The ratio of C-peptide to insulin was maximal when the concentration of both hormones was at its lowest. On day 55 (when the ratio was high), insulin binding and degradation were also increased. The decrease in the ratio by day 74 suggests a decrease in insulin binding and degradation by the tissues. By this stage, the expedition had become easier, and the intensification of secretion may reflect an increase in anabolic activity, an increase in lipogenesis, and a transfer to food that was richer in energy content.

The changes following the expedition (a return of glucagon to initial values, with subnormal values of insulin and C-peptide and an increased C-peptide/insulin ratio) may indicate that a moderate stress reaction accompanied reacclimatization, with activation of α-adrenergic mechanisms [Smith et al., 1979b] and an increased hepatic breakdown of insulin.

The data suggest that the activity of both α and β pancreatic cells was increased at days 29 and 74, but that, on day 55 and following the expedition, the α cells were operating at their initial level, while the β cells were depressed against the background of an increased insulin degradation in the liver.

Conclusions

Catecholamines and glucocorticoids activate catabolic processes to provide the stressed organism with appropriate substrates. In the early stages of a stress reaction, hyperglycemia and an increased level of free fatty acids can lead to a picture reminiscent of diabetes, with associated increases in blood lactate and ketones [Zaidase and Bessman, 1984]. However, the present expedition was marked by hypo- rather than hyperglycemia and by frank feelings of hunger, particularly at stages 2 and 3 of the trek.

Intense and prolonged physical strain leads to a decrease in blood levels of insulin, with an increased secretion of glucagon. Free fatty acid levels are increased as the glucose concentration falls, the consumption of both fatty acids and glucose by the muscles is increased, and hepatic gluconeogenesis is enhanced [Ahlborg et al., 1974; Galbo et al., 1979]. If cold is combined with physical exertion, these effects are intensified [Galbo et al., 1979; Murray et al., 1986]. Acclimatization to Siberia and the north is associated with decreased blood levels of insulin and glucose, and increased fat transport [Panin, 1987]. Starving and other kinds of stress lead to decreases in insulin levels [Udintsev et al., 1985; Jimenez et al., 1987], reflecting both a decrease in secretion and an increase in insulin degradation [Jimenez et al., 1987; Kosovskii et al., 1988].

Neuropeptide Concentrations

General Considerations

β-Endorphin is synthesized in the pituitary; production is enhanced by prolonged exercise and a rise of core temperature, and the substance contributes to the 'runner's high'. During periods of intense stress, ACTH and BE are secreted into the blood in equimolar amounts [Guillemin et al.,

Table 10.12. Blood levels of neurotransmitters and biologically active compounds (mean ± SD).

Location	Substance P pg/ml	cAMP pmol/l	cGMP pmol/l
Moscow	53.3 ± 16.3	14.9 ± 1.2	3.32 ± 0.27
Dikson	29.6 ± 2.4	11.0 ± 0.5[a]	4.43 ± 0.29[a]
Day 12	32.6 ± 3.7	33.6 ± 4.8[a,b]	1.62 ± 0.15[a,b]
Day 29	30.6 ± 3.1	35.4 ± 4.1[a,b]	1.08 ± 0.04[a,b]
Day 55	47.4 ± 4.5[b]	30.7 ± 1.9[a,b]	1.44 ± 0.07[a,b]
Day 74	40.6 ± 3.1[b]	9.4 ± 0.8[a]	1.22 ± 0.09[a,b]
Ottawa	44.3 ± 3.3[b]	7.7 ± 0.5[a,b]	4.56 ± 0.46[a]

[a,b] Significant differences relative to baseline tests conducted in Moscow (a) and data collected in Dikson (b).

1977]. However, an increase in this ratio can develop in chronic stress, and it leads to an increased pain sensitivity [Amir et al., 1980].

Substance P is a neuromediator for the primary sensory neurones. It not only transfers nociceptive impulses, but under some circumstances it can also suppress pain reactions [Fredrickson et al., 1978; Oehme et al., 1980a, b, 1985; Kryshanovskii et al., 1988].

Neurotensin is concentrated in hypothalamic structures and in the intestines. It induces vasodilatation, and thus hypotension. The increase of blood content is influenced by patterns of nutrition [Mashford et al., 1978].

Changes during the Expedition

Blood BE concentrations decreased significantly on days 29 and 74 of the expedition, and after its completion (table 10.5). Concentrations of substance P, on the other hand, increased considerably in the second half of the trek, and after completion of the mission (table 10.12). Neurotensin levels increased at stages 3 and 4 of the trek (table 10.12).

The ratio ACTH/BE increased progressively, with maximal values at stage 4 (day 74) and after the traverse (table 10.5, 10.7, 10.12); this is the anomalous pattern normally associated with an increase in pain sensitivity. Since there was no increase in BE, any biochemical explanation for a decreased pain sensitivity as the expedition proceeded must reflect activation of the antinociceptive system by substance P [Kryshanovskii et al., 1988], perhaps related to the binding of substance P by various receptors [Oehme et al., 1978, 1980a, b; Kryshanovskii et al., 1988]. Substance P

PGE ng/ml	Thromboxane pg/ml	Prostacycline (6-keto-$PGF_{1\alpha}$) pg/ml	Neurotensin pg/ml
236.0 ± 30.5	456.3 ± 163.8	128.3 ± 20.5	31.1 ± 3.7
240.7 ± 16.8	1167.2 ± 157.8[a]	187.9 ± 25.9	36.2 ± 2.5
247.5 ± 17.5	3075.0 ± 254.9[a,b]	144.0 ± 23.1	29.6 ± 5.5
500.0 ± 67.3[a,b]	2850.6 ± 437.3[a,b]	255.8 ± 39.3[a]	34.6 ± 5.3
45.3 ± 5.56[a,b]	1630.6 ± 108.7[a,b]	194.5 ± 16.5[a]	48.5 ± 5.0[a,b]
329.5 ± 38.9	2666.4 ± 243.0[a,b]	295.1 ± 26.5[a,b]	49.0 ± 4.0[a,b]
126.2 ± 13.2[a,b]	1397.6 ± 207.0[a]	212.7 ± 14.0[a]	31.3 ± 4.6

further seems able to decrease the neurotic states [Geht et al., 1988] and normalize sleep, improving memory and training [Losev et al., 1988]. It is thus an important modulator of both physiologic and pathologic processes [Oehme et al., 1980a, b].

During the first half of the trek, there was an increase in blood serotonin (above), which apparently was due to the increased motor activity of the skiers; serotonin is known as a motor mediator [Banister and Griffiths, 1972]. The serotoninergic system, together with the endogenous opioid peptides and substance P, is involved in regulation of the antinociceptive brain system, weakening pain perception. In the first part of the expedition, an increased ACTH/BE ratio testified to an increase of pain sensitivity, a fact also noted by the expedition's physician. The level of substance P was unchanged, and blood serotonin (which decreases pain sensitivity) was increased. Therefore, the increase in pain sensitivity at this stage was predominantly mediated by altered relative levels of ACTH and BE.

During the second half of the expedition, there was a gradual decrease in blood serotonin, this being especially marked at the end of the expedition. The ratio ACTH/BE increased still more, and the blood content of substance P significantly increased. The expedition's physician reported a decreased sensitivity to pain during this period. Given the changes of serotonin and ACTH/BE, both of which tended to increase pain sensitivity, the decrease of pain sensitivity observed by the expedition's physician must have been influenced predominantly by the increased concentration of substance P (which decreases pain sensitivity).

Therefore, regulation of the antinociceptive and nociceptive brain systems apparently changed during the course of the expedition. In the first half of the trek, the dominant role in activation of the nociceptive system was played by the blood ACTH/BE ratio, but in the second half activation of the antinociceptive system, stimulated by substance P, prevailed.

Biologically Active Compounds

General Considerations

cAMP is the mediator for adrenaline and glucagon, and thus acts at several points in the metabolic chain to modify the relative proportions of carbohydrate and fat metabolism.

Prostaglandins have a multiplicity of functions. They inhibit adenylate cyclase, and possibly reduce the release of NA from nerve terminals. Cycling to exhaustion causes the release of some prostaglandins, contributing to an increase in both skin and muscle blood flow. They also increase myocardial contractility, decrease fat mobilization, and regulate ion transport across some membranes.

Changes during the Expedition

Blood levels of cAMP increased sharply over the first three stages of the expedition, with the highest levels being observed on day 29 (table 10.12). By stage 4 of the traverse, concentrations had returned to their initial values, and levels were even lower after the expedition. Urinary excretion of cAMP decreased only on day 12 (table 10.6). Blood levels of cGMP decreased throughout the expedition (table 10.12), and excretion of cGMP was also depressed (table 10.6). The ratio cAMP/cGMP thus increased steeply throughout the trek (table 10.7], and the urinary cAMP/cGMP ratio also significantly exceeded initial values. The increase in cAMP/cGMP at stage 1 suggests the first response to a strain on the adrenergic system [Kozhemyakin et al., 1977]. At stage 2 of the traverse, there was a sharp (12-fold) increase in cAMP/cGMP ratio, indicating the growing strain on the adrenergic system. By stage 3, cAMP levels were sustained, but the cAMP/cGMP ratio decreased somewhat; possibly, cholinergic regulation was becoming more important at this stage. A decrease in blood levels of ACTH and glucagon may also have contributed to the decrease in cAMP, since the secretion of these hormones is regulated via adenylate cyclase-cAMP [Chowers et al., 1976]. By stage 4 of the expedition, cAMP had returned to initial values, although cGMP remained below the initial readings, suggesting further development of cholinergic regulation. After the expedition, cAMP levels were lower than initially, suggesting a decrease in sympathoadrenal activity.

Increased blood levels of PGE at stage 2 of the trek (table 10.12) probably reflect the influence of the sympathoadrenal system upon the synthesis of PGE [Nebolsina et al., 1982; Meerson, 1984]. PGE was markedly decreased by day 55, 11-fold relative to stage 2 of the trek, and 5-fold relative to initial values in both Moscow and Dikson. These changes are probably secondary to the decrease in cAMP, since this substance activates the phospholipase A_2 involved in prostaglandin synthesis [Mevkh, 1983]. An increase in lipolysis and the release of lipids into the blood stream could also contribute to the decrease in PGE [Bochkareva, 1980; Bergelson, 1986]. The increase in lipolysis is implicit in the increased blood levels of GH observed at this stage. The increase in prostaglandin at stage 4 of the traverse may be related to hyperventilation, with promotion of prostaglandin synthesis [Said, 1981]. An alteration of diet at this stage (particularly the inclusion of polyunsaturated fatty acids) could also have contributed to increased prostaglandin readings. Following the expedition, PGE levels were below initial values, perhaps due to a reduction in sympathetic activity.

The blood concentration of 6-keto-($PGF1_\alpha$) (prostacycline) was considerably increased at days 29, 55 and 74, and following the mission, with a corresponding tendency to vasodilatation (table 10.12). Blood levels of thromboxane were increased throughout (table 10.12). The increase in thromboxane against a background of decreasing prostacycline is characteristic of the adjustment of the body to new conditions. When the sympathoadrenal system is first activated, thrombocyte aggregation is increased, with an enhanced synthesis of thromboxane, and a decrease in prostacycline [Aliev et al., 1984]. The final concentrations of thromboxane did not differ from the initial values.

Urinary excretion of both cAMP and cGMP was high in Dikson, suggesting a stress reaction. The output of cGMP was decreased throughout the expedition, and urinary elimination of cAMP was also decreased at stage 1, probably due to an increased utilization of both compounds in the tissues (table 10.6).

Stress, Carbohydrate and Lipid Metabolism

General Considerations
Metabolic adjustments to stress may be either acute or chronic. The acute reactions are intended to mobilize substrates (glucose and free fatty acids), while chronic adjustments are intended to spare intramuscular stores of glycogen through the preferential oxidation of lipids.

The acute reaction of carbohydrate and lipid mobilization depends on the actions of glucagon and adrenaline upon glycolytic enzymes in the liver and muscles, and upon the adenylate cyclase-cAMP system in depot fat.

The chronic changes reflect in part altered tissue enzyme activity, and in part an altered balance of cortisol and insulin. Cortisol acts mainly by increasing the activity of hepatic enzymes involved in amino acid metabolism, gluconeogenesis and lipolysis. Insulin acts upon hexokinase, favoring transport of glucose into muscle and liver to replenish glycogen depots. It also suppresses the enzymes involved in gluconeogenesis, it reduces lipolysis, and promotes the synthesis of proteins. There are three potential manifestations of the chronic stress reaction, depending on the severity of the situation: (1) there may be an increase in blood glucocorticoids without any change in insulin level; (2) there may be an unaltered glucocorticoid level but a decrease in insulin, or (3) there may be a simultaneous decrease in glucocorticoids and a decrease in insulin concentration.

Changes during the Expedition

Blood glucose was decreased during and following the expedition (table 10.1). Serum cholesterol levels showed no substantial changes, but the total blood lipids were increased at stages 2 and 4. Triglyceride levels had decreased at day 55, but were sharply increased at day 74 (table 10.1).

The changes in serum lipid profile are somewhat ambiguous, but seem to mirror parallel changes in the lipid content of the tissues [Alimova et al., 1975]. There are increases in total lipids, free fatty acids and very low-density lipoproteins [chapter 10], these changes reflecting both an increased mobilization of fat and an enhanced formation of transport lipids in the liver [Panin, 1978]. The shift to lipid metabolism [Panin, 1970; Karlson et al., 1970; Vlasova et al., 1975] is mediated by a number of hormones, including GH, cortisol, insulin and glucagon [Panin, 1970], the first two of these increasing during the expedition (above). During the second half of the expedition, lipogenic processes begin to prevail, with increased blood levels of triglyceride and insulin, and decreasing levels of cortisol.

In the first stage of the expedition (to day 29), there was a pronounced decrease in blood glucose, and during the remainder of the trek glucose levels seemingly stabilized at this lower level. These changes may be related to the previously discussed modifications in insulin and glucagon levels. In particular, an increase in insulin accelerates the transport of glucose into the liver and muscles and enhances peripheral glucose utilization by the tissues. In contrast, glucagon enhances hepatic glycolysis and gluconeogenesis.

Thus, the stabilization of blood glucose levels reflects the striking of a new balance between the blood levels of insulin and glucagon. The final glucose readings were increased relative to day 74 of the trek, coincident with decreases in insulin and glucagon.

Metabolic studies of migrants to high latitudes have shown a decrease in anaerobic processes in the erythrocytes, with a similar tendency to hypoglycemia [Panin, 1978]. However, the low blood glucose does not seem to be accompanied by any of the usual objective signs of hypoglycemia. It seems that carbohydrates assume a diminished role in the provision of energy to polar explorers [Panin, 1983].

Evidence of Stress and Adaptive Responses

The development of a stress reaction during the expedition was suggested by increased blood levels of catecholamines and glucocorticoids, decreases of gonadotrophic and gonadal activity, increases in GH and parathyroid hormone secretion, and increased levels of vasopressin, angiotensin-II, aldosterone, cAMP, TSH, and plasma renin activity.

Stress mechanisms involve the autonomic system, via the hypothalamus and centers in the medulla oblongata. These centers activate the sympatho-adrenal system, the hypophyseal adrenocortical system, and various respiratory, cardiovascular and renal processes. Metabolic processes are also modified via the endocrine responses.

The initial, acute, hormonal reactions are an increased output of adrenaline and glucocorticoids from the medulla and cortex of the adrenal glands, respectively. Activation of the neurohypophysis also increases the output of vasopressin. The consequences are an increase in glycolysis and hepatic gluconeogenesis, a release of erythrocytes from the spleen, a decrease in diuresis, and an increase in arterial pressure. At this stage, the depressor effect of the aortic baroreceptors is augmented, restoring homeostasis. Commonly, stimuli from internal receptors are supplemented with emotionally colored information derived from the limbic system. This can augment or correct any autonomic imbalance.

The chronic reaction to stress is a sustained activity of the sympathoadrenal system and the hypophysis/adrenocortical system. The resultant hormonal changes lead to psychosomatic disorders in association with anxiety and neuroticism [Panin, 1983]. Nevertheless, stress is not always accompanied by manifest anxiety.

The increases in catecholamine that we observed seem a true expression of stress, rather than an alteration in catecholamine metabolism, since no changes in synthesis or inactivation of catecholamines were seen (with the possible exception of dopamine, above). The increased blood A/NA ratio confirmed that there was an emotional stress reaction at stage 1 of the expedition. The cortisol levels were biphasic, with increases relative to ACTH in the first half of the mission, but not in the second half. As already noted, this may reflect problems of hydroxylation (at atom 17 of the molecule, above). The persisting decrease in the ACTH/cortisol ratio suggests a chronic stressing of this axis [Dilman et al., 1987].

Mineralocorticoid secretion was apparently enhanced throughout the expedition, with blood levels being stabilized after stage 1 by an increase in urinary excretion.

Finally, we may note the role of insulin, which normally serves to counteract the metabolic effects of cortisol. In the first half of the expedition, the increased glucocorticoid output was matched by increased activity of the β cells of the pancreas, but in the second half of the trek cortisol levels were less high, and insulin secretion was also less marked.

11 Comparison of Biochemical Reactions to Trek and Chamber Simulations[1]

L.E. Panin, N.M. Mayaskaya, A.A. Borodin, P.E. Vloshinsky,
I.E. Kolosova, A.R. Kolpakov, V.G. Kunitsin, M.F. Nekrasova,
N.G. Kolosova, L.S. Ostanina, T.A. Tretyakova

Introduction

Biochemical observations on both metabolites and hormones obtained during the trek were compared to results observed during chamber simulations of the combined cold exposure and exercise stress [chapter 7], as undertaken in Moscow (before the expedition) and in Ottawa (after completion of the mission). During the climatic chamber experiments, peripheral venous blood samples were collected from the antecubital vein initially, after a night in the cold chamber, and after 3 h of stepping exercise.

Data were obtained on glucose-regulating hormones, serum proteins, blood lipid profiles, lipid peroxidation, erythrocyte glycolysis, acid lysozymes, and several measures of erythrocyte membrane properties (viscosity, electrical conductivity, and infrared spectroscopy).

Glucose-Regulating Hormones

Methodology
Urinary glucose concentrations were assayed by the orthotoluidine method, spectrophotometric measurements being made at 620 nm. All hormones were determined by radioimmune techniques.

Responses before the Expedition
When first tested in Moscow, adrenocorticotropin (ACTH) and cortisol levels tended to decrease while sleeping in the chamber, and to recover towards normal values during the exercise bouts (table 11.1). However, the high cortisol readings were anomalous relative to the ACTH values, which were below the laboratory norms of 63.0 ± 9.0 ng/l, suggesting that factors other than ACTH may have influenced blood cortisol readings at this stage.

[1] Freely adapted from the Russian translation by R.J.S.

Table 11.1. Hormone concentrations during chamber simulations of the expedition before (Moscow) and after (Ottawa) the ski trek

Hormones	Moscow 1	Moscow 2	Moscow 3	Ottawa 1	Ottawa 2	Ottawa 3
ACTH, pg/ml	41.1 ± 8.9 (n = 3)	29.7 ± 9.6 (n = 3)	40.5 ± 5.5 (n = 9)	40.4 ± 4.5 (n = 13)	57.9 ± 3.0 (n = 9)	43.6 ± 7.7 (n = 7)
Cortisol nmol/l	565.0 ± 78.0 (n = 9)	346.0 ± 68.1 (n = 8)	522.0 ± 58.1 (n = 11)	927.0 ± 70.0** (n = 13)	973.0 ± 52.0** (n = 11)	655.0 ± 53.0 (n = 10)
Insulin µU/ml	7.90 ± 0.35 (n = 6)	10.80 ± 0.69* (n = 6)	14.70 ± 2.60* (n = 11)	9.10 ± 1.10 (n = 13)	11.00 ± 0.70 (n = 10)	6.90 ± 1.10** (n = 9)
C-peptide ng/ml	0.49 ± 0.05 (n = 10)	0.60 ± 0.07 (n = 6)	0.71 ± 0.07* (n = 11)	1.24 ± 0.09** (n = 13)	1.40 ± 0.06** (n = 10)	1.06 ± 0.06** (n = 9)
Glucagon pg/ml	43.2 ± 4.4 (n = 5)	38.8 ± 6.4 (n = 5)	60.2 ± 6.7 (n = 10)	51.0 ± 8.8 (n = 12)	47.5 ± 10.0 (n = 11)	39.1 ± 9.5 (n = 10)
TSH mIU/l	1.56 ± 0.25 (n = 6)	2.60 ± 0.54 (n = 6)	2.60 ± 0.38* (n = 11)	2.68 ± 0.50 (n = 11)	3.43 ± 0.60 (n = 10)	3.13 ± 0.80 (n = 8)
T_3, nmol/l	1.20 ± 0.06 (n = 6)	1.20 ± 0.07 (n = 6)	1.30 ± 0.06 (n = 11)	1.52 ± 0.06** (n = 13)	1.59 ± 0.06** (n = 10)	1.56 ± 0.08** (n = 10)
T_4, nmol/l	83.7 ± 6.0 (n = 6)	79.0 ± 2.7 (n = 6)	90.6 ± 3.5 (n = 11)	89.1 ± 4.0 (n = 13)	89.2 ± 5.4 (n = 10)	95.0 ± 4.1 (n = 10)
Aldosterone pg/ml	101.0 ± 10.2 (n = 9)	86.6 ± 10.8 (n = 8)	148.1 ± 16.6* (n = 11)	78.3 ± 10.2 (n = 9)	107.0 ± 14.2 (n = 10)	89.0 ± 11.5** (n = 9)
Growth hormone ng/ml	0.54 ± 0.14 (n = 7)	0.57 ± 0.06 (n = 7)	2.39 ± 0.85* (n = 10)	0.48 ± 0.07 (n = 8)	0.71 ± 0.08 (n = 8)	0.73 ± 0.12** (n = 5)

1 = Basal metabolism; 2 = sleep in the cold chamber; 3 = cold chamber + exercises. Differences ($p < 0.05$): *only in Moscow (1–2, 1–3 accordingly); **with Moscow (1–1, 2–2, 3–3 accordingly).

Blood insulin concentrations were increased by cold exposure, and were further increased by the subsequent exercise (table 11.1). Since there was a parallel increase in the precursor metabolite c-peptide, this suggests an increase in pancreatic activity; possibly, there was a general pancreatic secretory reaction to increased blood glucose concentrations, since glucagon concentrations were also markedly increased by the exercise.

Sleep in the cold chamber did not change the blood levels of the thyroid hormones significantly (table 11.1), although exercise was associated with a significant increase in thyroid-stimulating hormone (TSH), and a trend to an increase in thyroxine (T_4). The exercise bouts induced significant increments in aldosterone and growth hormone (table 11.1).

Almost all these reactions may be considered as hormonal adjustments designed to service the increased metabolism. The initial association of an increase in glucocorticoids with a decrease in blood insulin levels points to the third type of chronic stress reaction during preparations for the trek [chapter 10]. However, when the sympathetic reaction to an acute cold stress was superimposed upon the chronic reaction to preparations for the expedition, cortisol decreased (the more favorable type-2 reaction).

Responses after the Expedition

Immediately following the expedition, the resting data collected in Ottawa showed further large increments in resting cortisol levels, again without a matching increment in ACTH.

In contrast to the initial reactions before the expedition, the exercise bout now induced a statistically significant decrement in insulin levels, a condition that the Soviet authors term 'strain diabetes'. However, c-peptide levels were increased, suggesting that there had been no decrease in pancreatic secretion. Possibly, the rate of breakdown of insulin by insulinase was increased. Strain diabetes is characterized by a decreased blood glucose, a decreased renal threshold for glucose and glycosuria. It seems to be associated with a switch from carbohydrate to fat metabolism under conditions of chronic stress. There may also be a decrease in blood glucagon levels, as in the present instance.

Relative to the data obtained in Moscow, thyroid function was somewhat enhanced in Ottawa, which might be interpreted as a response to the trek; however, if the Ottawa findings are compared with those for the general population, it would appear that the discrepancy is due mainly to low levels of thyroid activity in Moscow.

In contrast with the data obtained in Moscow, the cold chamber experiments induced no changes in glucagon, T_4, aldosterone or growth hormone concentration.

Thus, the overall baseline data observed in Ottawa could again be considered as evidence of the third type of reaction to chronic stress [chapter 10]. While the night of sleeping in the cold had now become a habitual event, with little influence upon hormone levels, the addition of a physical load induced decreases in cortisol, insulin and c-peptide, with ACTH remaining constant; this is an example of the second, most favorable pattern of reaction to stress [chapter 10]. The evidence of a high baseline of stress, after the trek had been completed, appears at first inspection somewhat paradoxical. However, it may be that there was, at this stage of the mission, a discharge of some of the negative emotions that had been restrained during the traverse.

Serum and Urinary Proteins

Methodology
Serum and urinary proteins were determined by the method of Lowry et al. [1951], in which the spectrophotometric absorbance of copper derivatives of the proteins is detected at a wavelength of 750 nm. Protein fractions were separated by paper electrophoresis, using a veronal acetate buffer; the paper strips were processed with bromphenol blue, and the color of the eluates was determined by spectrophotometer at a wavelength of 590 nm.

Changes over the Expedition
The initial values for total serum proteins averaged 76.8 g/l in Moscow and 83.7 g/l in Dikson, compared with an average range of 80–86 g/l commonly reported for the Moscow population (table 11.2). The increase in concentrations from Moscow to Dikson Island was attributable to an augmentation of albumins and α_1-globulins. The total protein and albumin levels diminished significantly over the course of the traverse. However, once in Ottawa, normal values were quickly restored. The level of α_2-globulins remained elevated throughout the expedition, but the basis for this change remains unclear.

Similar changes were observed previously when northern residents were compared with the general population. The explanation seems to be a glucocorticoid-induced increase in gluconeogenesis, with albumins serving as the main source of the additional amino acids to be metabolized. Since the present group did not develop a gross hypoproteinemia or hypoalbuminemia over the trek, it suggests that their protein intake was sufficient to compensate at least partially for the enhanced gluconeogenesis. The albumin/globulin ratio remained within the normal range for the Soviet population (1.2–1.8) throughout the mission.

Table 11.2. Changes in serum proteins (g/l) during the course of the ski trek

	Moscow	Dikson	Stage I	Stage II	Stage III	Ottawa
Total protein	76.8 ± 2.6 (n = 7)	83.7 ± 3.1 (n = 7)	70.5 ± 3.9* (n = 7)	73.1 ± 1.7* (n = 7)	74.1 ± 4.5 (n = 7)	78.5 ± 3.4 (n = 7)
Albumins	47.2 ± 1.8 (n = 6)	52.8 ± 1.9 (n = 7)	42.8 ± 2.5* (n = 7)	46.3 ± 1.7* (n = 7)	46.5 ± 2.9 (n = 7)	48.1 ± 2.1 (n = 7)
α_1-globulins	3.0 ± 0.2 (n = 6)	3.1 ± 0.2 (n = 7)	3.1 ± 0.3 (n = 7)	2.9 ± 0.2 (n = 7)	2.8 ± 0.2 (n = 7)	3.2 ± 0.3 (n = 7)
α_2-globulins	4.3 ± 0.2 (n = 6)	5.4 ± 0.4* (n = 7)	5.2 ± 0.3* (n = 7)	5.1 ± 0.2* (n = 7)	5.2 ± 0.4 (n = 7)	4.8 ± 0.3 (n = 7)
β-globulins	7.1 ± 0.3 (n = 6)	7.9 ± 0.5 (n = 7)	6.7 ± 0.5 (n = 7)	6.8 ± 0.2 (n = 7)	7.2 ± 0.6 (n = 7)	7.7 ± 0.5 (n = 7)
γ-globulins	15.1 ± 1.7 (n = 6)	14.5 ± 1.3 (n = 7)	12.7 ± 0.8 (n = 7)	11.9 ± 0.2 (n = 7)	12.4 ± 1.1 (n = 7)	14.6 ± 1.9 (n = 7)

*Differences (p < 0.05) from initial (Moscow) data.

Responses before the Expedition

During the initial climatic chamber experiments (table 11.3), the combination of cold and exercise induced a slight increase in total protein, serum albumin and γ-globulins. This was thought due to dehydration and a resultant decrease in plasma volume.

Responses after the Expedition

These reactions were not seen when the same experiments were repeated in Ottawa after the expedition, the difference between the two chamber experiments being statistically significant (table 11.3).

Blood Lipid Profile

Methodology, Initial Norms and Overall Fat Content of Blood

Analysis of the blood lipid profile included determinations of free fatty acids (FFAs), triglycerides, phospholipids, the sum of very low-density lipoprotein (VLDL) and low-density lipoprotein (LDL), cholesterol, free cholesterol, high-density lipoprotein (HDL) cholesterol, changes in lipoprotein patterns and in the spectrum of phospholipids.

FFAs were estimated by the method of Dumcombe [1964], using a copper reagent (cuprous nitrite, triethanolamine and acetic acid) with

Table 11.3. Changes in serum proteins (g/l) during climatic chamber experiments before (Moscow) and after (Ottawa) the ski trek

	Moscow			Ottawa		
	1	2	3	1	2	3
Total protein	76.8 ± 2.6 (n = 7)	85.9 ± 3.6 (n = 7)	85.0 ± 2.5* (n = 7)	78.5 ± 3.4 (n = 7)	72.5 ± 2.6** (n = 7)	77.6 ± 2.2** (n = 7)
Albumins	47.2 ± 1.8 (n = 6)	53.4 ± 2.2* (n = 7)	51.5 ± 1.0 (n = 7)	48.1 ± 2.1 (n = 7)	44.5 ± 1.9** (n = 7)	46.4 ± 1.2** (n = 7)
a_1-globulins	3.0 ± 0.2 (n = 6)	3.2 ± 0.4 (n = 7)	3.2 ± 0.2 (n = 7)	3.2 ± 0.3 (n = 7)	3.0 ± 0.1 (n = 7)	3.3 ± 0.3 (n = 7)
α_2-globulins	4.3 ± 0.2 (n = 6)	5.2 ± 0.5 (n = 7)	4.8 ± 0.3 (n = 7)	4.8 ± 0.3 (n = 7)	4.7 ± 0.4 (n = 7)	4.9 ± 0.3 (n = 7)
β-globulins	7.1 ± 0.3 (n = 6)	7.7 ± 0.5 (n = 7)	7.9 ± 0.4* (n = 7)	7.7 ± 0.5 (n = 7)	6.9 ± 0.4 (n = 7)	7.8 ± 0.4 (n = 7)
γ-globulins	15.1 ± 1.7 (n = 6)	16.3 ± 2.1 (n = 7)	17.6 ± 1.3* (n = 7)	14.6 ± 1.9 (n = 7)	13.4 ± 1.3 (n = 7)	15.3 ± 1.4 (n = 7)

1 = Basal measurements; 2 = sleep in the cold chamber; 3 = cold chamber + exercises.
Differences ($p < 0.05$): *only in Moscow (1–2, 1–3 accordingle); **with Moscow (1–1, 2–2, 3–3 accordingly).

sodium diethylcarbamate in butanol. The absorbance was calculated relative to standard preparations of palmitic acid.

Serum triglycerides were determined with LAHEMA test reagents, which in essence used calcium hydroxide to saponify the fat, and then release formaldehyde by oxidation. The formaldehyde reacted with methyl acetone and ammonium ions to form a yellow substance, 3,5-diacetyl-1,4-dihydrolutidine, that was assayed spectrophotometrically.

Total serum lipids were determined by interaction with phosphovanillic acid after sulfuric acid hydrolysis, the resultant red color being assayed by spectrophotometer.

The sum of VLDL and LDL particles was assessed by turbidimetry, as the precipitate formed with heparin in the presence of calcium ions.

Serum lipoproteins were fractionated by polyacrylamide gel electrophoresis, with subsequent Sudan black-B treatment. This technique allowed the assay of VLDL, LDL, HDL_2, HDL_3 and complexes of albumin with the fatty acids, the results of spectrophotometric assay of the eluate correlating well with the ultracentrifugation technique of Hatch and Lees [1968].

Phospholipids were measured by thin-layer chromatography on silufol discs, with subsequent scanning by Hitachi spectrofluorimeter.

Table 11.4. Changes in lipid profile during the course of the ski trek

Lipid fraction	Moscow	Dikson	Stage I	Stage II	Stage III	Ottawa
ГГА, mmol/l	181 ± 22.7 (n = 9)	172 ± 13.7 (n = 13)	188 ± 36.9 (n = 7)	208 ± 14.4 (n = 8)	160 ± 11.3 (n = 11)	170 ± 9.3 (n = 13)
Triglycerides mg/100 ml	107 ± 14.6 (n = 8)	88 ± 6.58 (n = 13)	83 ± 9.97 (n = 7)	98 ± 16.30 (n = 9)	70 ± 3.42* (n = 11)	93 ± 7.75* (n = 13)
LDL + VLDL mg/100 ml	644 ± 37.0 (n = 10)	701 ± 45.3 (n = 13)	479 ± 46.6** (n = 10)	473 ± 46.6** (n = 8)	526 ± 48.6* (n = 11)	553 ± 31.2* (n = 13)
HDL Cholesterol mg/100 ml	53.0 ± 2.7 (n = 8)	64.6 ± 2.0* (n = 13)	74.4 ± 3.1** (n = 8)	83.8 ± 4.6** (n = 8)	80.6 ± 4.2** (n = 12)	83.1 ± 3.6** (n = 11)
Total lipids mg/100 ml	5.63 ± 0.19 (n = 9)	5.71 ± 0.99 (n = 9)	4.06 ± 0.16 (n = 10)	4.68 ± 0.34 (n = 13)	4.83 ± 0.94 (n = 10)	5.63 ± 0.19 (n = 9)

Difference ($p < 0.05$) with: *Moscow; **Dikson and Moscow.

The baseline values, and laboratory norms (in parentheses) for these variables, based upon the general Moscow population, were:

FFAs 327 ± 24 (300–350) nmol/l
Triglycerides 1.0 ± 0.1 (0.8–1.0) g/l
Phospholipids 2.0 ± 0.07 (1.5–2.5) g/l
VLDL + LDL cholesterol 6.4 ± 0.3 (4.0–6.0) g/l
HDL cholesterol 0.52 ± 0.02 (0.5–0.6) g/l.

Changes in FFAs and Triglycerides over the Expedition

The initial values observed for the Soviet team members were essentially normal, except that the data collected in Moscow and Dikson before the expedition showed low FFA readings (table 11.4). The half-life of FFAs is extremely short (2–4 min), and it may be that, in the present well-trained group, there was a particularly high metabolic rate for FFAs even before the expedition began. Certainly, the low FFA readings do not seem to have been particularly harmful, as they did not change over the trek, or subsequently in Ottawa.

Triglyceride levels change much more slowly. Although there was a downward trend during the trek, the only significant decrement was seen at stage 3, corresponding to samples collected at the North Pole. On reaching Ottawa, values returned to their initial level.

Serum Lipid Profile

Serum lipoproteins are of great interest in metabolism, since they are the active transport portion of total lipids.

Theoretical Considerations

The largest particles (chylomicrons) have a molecular mass of $0.5-4.3 \times 10^9$. They comprise 84–90% triglycerides, 4.3–7.0% phospholipids, 4.0–5.0% cholesterol esters, 2.0% free cholesterol, and 2.0% proteins. Blood concentrations range from 0–0.5 g/l, and their half-life in the blood stream is 5–15 min. The main function of these particles is the transport of lipid to the heart and fatty tissues, where they are partly destroyed by the lipoprotein lipase (LPL) bound to the capillary endothelium. The products of the LPL activity (the 'remnants' of the chylomicrons) are consumed by the liver. We did not take chylomicrons into account in this study, since they have to be removed from the blood before the determination of LDL and VLDL particles, and they are not displaced during polyacrylamide electrophoresis.

VLDL particles have a molecular mass of $3-10 \times 10^6 D$. They contain 55–65% triglycerides, 13–20% phospholipids, 10–13% cholesterol esters, 3–5% free cholesterol, and 5–12% proteins. The blood concentrations in adults range from 0.93–0.97 g/l, and the half-life of the particles is about 40 min. They are broken down by LPL in the capillaries of the peripheral tissues. The liberated FFA is oxidized locally, while the apoproteins pass into part of the HDL particles. The phospholipids and cholesterol from the VLDL are also thought to shift from the remnants of the particle to HDL, the process being promoted by the enzyme lecithin-cholesterol-acyltransferase (LCAT). As a result, some 72–83% of the VLDL is transmuted into HDL, and the remainder is metabolized by the liver without transformation. Triglycerides bound to the VLDL are consumed intensively by the heart and skeletal muscle, particularly during vigorous and prolonged exercise. Physical activity thus has the effect of decreasing blood levels of VLDL and increasing HDL concentrations.

The LDL particles have a molecular mass of $2.1-2.6 \times 10^6 D$. The composition includes 10–12% triglycerides, 20–22% phospholipids, 36–37% cholesterol esters, 8–10% free cholesterol, and 21–25% proteins. Blood levels are typically 3.69 ± 0.81 g/l, and the half-life is 3–5 days. Catabolism occurs in the liver, the adrenal glands, spleen and ovaries, with other tissues consuming no more than 8% of the total blood pool. The functional significance of LDL is not completely known, but it seems to be the main carrier of cholesterol in the body.

HDL particles typically have a mass of $0.2 \times 10^6 D$. They contain 3–7% triglycerides, 27–30% phospholipids, 14–20% cholesterol esters, 2–4% free cholesterol and 45–55% proteins. The method of centrifugation available to the Soviet investigators distinguished two HDL subfractions with specific gravities in the range 1.063–1.125 and 1.125–1.210 g/ml, respectively. The first fraction carried 60% lipids and 40% proteins, while the second contained 45% lipids and 55% proteins. HDL particles are formed in part from VLDL particles in the capillaries. The half-life of HDL is 4–5 days. The functional significance of HDL is again not entirely clear, but it does serve to transport cholesterol from peripheral tissues to the liver, where it can be transformed into bile acids and excreted. In the plasma, LPL may convert larger to the smaller HDL particles. During the process of conversion, HDL_2 loses some cholesterol esters and triglycerides, while HDL_3 takes up cholesterol from the tissues and is transformed into HDL_2 under the influence of LCAT.

Serum Lipid Profile during the Trek

The initial VLDL + LDL levels observed in Moscow were normal (table 11.4), and there was an insignificant trend to an increase at Dikson, with a subsequent 25–30% decrease over the traverse. Presumably, these two fractions of the total lipid were serving as an important source of FFAs during the expedition. The associated transfer of cholesterol from VLDL and cell membranes to HDL led to a substantial and progressive increase in HDL cholesterol, 21% at Dikson, 40% after 2 weeks of skiing, and 58% after 4 weeks of skiing.

The ratio of HDL/LDL is often considered in the context of coronary risk, but also indicates the relative proportions of lipid and carbohydrate metabolism. In Moscow, the ratio of HDL/ (VLDL + LDL) was 1.0, as compared with the Moscow norm of 0.4–0.5. This suggests that even at the outset of the expedition (as might be expected in endurance-trained subjects), lipids were making an above average contribution to metabolism. Over the course of the trek, the ratio rose to 2.0, as HDL increased and VLDL + LDL decreased. The data suggest a further increase in LCAT activity, and a further decrease in the risk of atherosclerosis under the conditions of the trek.

Total lipids showed an insignificant change from Moscow to Dikson, with a significant decrease after 2 weeks of skiing, and a subsequent gradual increase (table 11.4).

There were dramatic decreases in both VLDL (from 1.17% in Moscow to 0.49% at the North Pole; (table 11.5) and in LDL (from 48.5% to 24.8%). The overall increase in HDL was associated with increases in the ratio of HDL_2 to HDL_3, from 0.9 in Moscow to 1.47 after 4 weeks of the traverse (table 11.5). The shift towards the HDL_2 fraction is again a favorable change in the lipid profile, and is probably due to increased LCAT and LPL activity. Previous observations have shown increases in the activity of LPL and heparin-independent triglyceride lipase among those living and working in the far north; in addition to the high work rates expected of many northern residents, the shift to fat metabolism may help

Table 11.5. Changes (%) in lipoprotein profile during the course of the ski trek

Lipoprotein fractions	Moscow	Dikson	Stage I	Stage II	Stage III	Ottawa
HDL_3	26.20 ± 1.65 (n = 10)	26.10 ± 2.44 (n = 13)	32.50 ± 2.14* (n = 11)	31.40 ± 2.15 (n = 9)	34.20 ± 2.14* (n = 11)	27.50 ± 1.56 (n = 13)
HDL_2	24.10 ± 1.96 (n = 10)	28.00 ± 2.25 (n = 13)	42.80 ± 1.71* (n = 11)	46.20 ± 2.80* (n = 9)	40.50 ± 2.49* (n = 11)	34.80 ± 3.11 (n = 13)
LDL	48.50 ± 4.18 (n = 10)	45.40 ± 2.80 (n = 13)	24.00 ± 1.87* (n = 11)	22.30 ± 3.14* (n = 9)	24.80 ± 3.32* (n = 11)	36.20 ± 3.47 (n = 13)
VLDL	1.17 ± 0.54 (n = 10)	0.56 ± 0.43 (n = 13)	0.68 ± 0.31 (n = 11)	0.11 ± 0.12 (n = 9)	0.49 ± 0.20 (n = 11)	0.73 ± 0.34 (n = 13)
$HDL_2 + HDL_3$	50.3	54.1	75.3	77.6	74.7	62.3
LDL + VLDL	49.7	45.9	24.7	22.4	25.3	37.7
$\frac{HDL}{(LDL + VLDL)}$	1.00	1.17	3.00	3.40	2.95	1.65

*Difference ($p < 0.05$) from the data obtained in Moscow.

Table 11.6. Changes in the phospholipid spectrum of the blood sera (mg/100 ml) during the course of the ski trek

Phospho-lipids	Moscow	Dikson	Stage I	Stage II	Stage III	Ottawa
PHH	82.2 ± 2.8	80.8 ± 2.0	75.2 ± 2.6	77.7 ± 2.1	84.0 ± 4.7	83.7 ± 1.2
PHEA	54.9 ± 1.4	55.9 ± 1.5	56.5 ± 1.2	52.9 ± 2.8	60.2 ± 3.3	62.4 ± 1.5
PHS	31.3 ± 3.5	32.7 ± 2.5	22.8 ± 3.1	27.0 ± 4.5	32.6 ± 1.7	32.7 ± 2.3
CL	10.2 ± 1.8	12.5 ± 0.9	10.3 ± 1.4	12.2 ± 0.9	12.1 ± 1.4	15.9 ± 1.1
S	26.2 ± 1.4	24.5 ± 0.9	22.5 ± 2.4	25.3 ± 2.7	21.8 ± 2.4	21.1 ± 1.4
LF	3.7 ± 1.3	6.9 ± 0.4	4.5 ± 0.5	6.0 ± 1.2	6.9 ± 0.9	4.6 ± 0.5
Total	208.7 ± 7.9	213.1 ± 8.4	191.8 ± 4.2	201.0 ± 5.4	217.6 ± 6.2	220.2 ± 7.3

PHH = Phosphatidylcholine; PHEA = phophatidylethanolamine; PHS = phosphatidylserine; CL = cardiolipin; S = sphingomyelin; LF = lysoforms.

to increase body heat production, because of the greater specific dynamic action of fat.

Phospholipid levels did not change over the course of the trek (table 11.6). The average turnover time of the phospholipids is about 3 days, and they are thus rather dormant indicators of the extent of lipid metabolism as compared to FFA or VLDL.

11 Comparison of Biochemical Reactions to Trek and Chamber Simulations

Table 11.7 Blood lipid profile during climatic chamber simulations of the expedition before (Moscow) and after (Ottawa) the ski trek

	Moscow			Ottawa		
	1	2	3	1	2	3
FFAs mmol/l	181.0 ± 22.7 (n = 9)	183.0 ± 23.0 (n = 7)	372.0 ± 43.5* (n = 11)	170.0 ± 9.25 (n = 13)	209.0 ± 30.6 (n = 10)	264.0 ± 35.7** (n = 7)
Triglycerides mg/100 ml	107 ± 14.6 (n = 8)	124 ± 19.9 (n = 7)	87 ± 10.1 (n = 10)	93 ± 7.7 (n = 13)	101 ± 6.7 (n = 11)	104 ± 12.4 (n = 7)
LDL + VLDL mg/100 ml	644 ± 37.0 (n = 10)	696 ± 51.4 (n = 7)	666 ± 50.4 (n = 11)	553 ± 31.2 (n = 13)	534 ± 28.0** (n = 10)	511 ± 43.0** (n = 7)
Total lipids mg/100 ml	5.04 ± 0.34 (n = 9)	5.64 ± 0.35 (n = 9)	5.32 ± 0.32 (n = 10)	5.63 ± 0.19 (n = 13)	4.37 ± 0.34** (n = 10)	4.84 ± 0.19** (n = 9)

1 = Basal measurements; 2 = sleep in the cold chamber; 3 = cold chamber + exercise.
Differences ($p < 0.05$): *only in Moscow (1–3); **with Moscow (2–2, 3–3).

Table 11.8. Blood lipoproteins (%) during climatic chamber simulations of the expedition before (Moscow) and after (Ottawa) the ski trek

Lipoprotein fraction	Moscow			Ottawa		
	1	2	3	1	2	3
HDL_3	26.2 ± 1.65 (n = 10)	28.2 ± 2.71 (n = 7)	24.0 ± 1.82 (n = 11)	27.5 ± 1.56 (n = 13)	29.5 ± 3.00 (n = 10)	29.2 ± 2.58 (n = 7)
HDL_2	24.1 ± 1.96 (n = 10)	24.8 ± 4.34 (n = 7)	26.7 ± 2.20 (n = 11)	34.8 ± 3.11** (n = 13)	33.6 ± 2.90 (n = 10)	37.6 ± 3.46** (n = 7)
LDL	48.5 ± 4.18 (n = 10)	46.7 ± 4.68 (n = 7)	47.6 ± 4.00 (n = 11)	36.2 ± 3.47** (n = 13)	35.6 ± 3.93 (n = 10)	32.2 ± 5.83** (n = 7)
VLDL	1.17 ± 0.54 (n = 10)	0.26 ± 0.26 (n = 7)	3.63 ± 0.60* (n = 11)	0.73 ± 0.34 (n = 13)	1.25 ± 0.54 (n = 10)	0.96 ± 0.61 (n = 7)
$HDL_3 + HDL_2$	50.3	53.0	50.7	62.3	63.1	66.8
LDL + VLDL	49.7	47.0	51.0	37.0	36.9	33.2
$\dfrac{HDL}{(VLDL + LDL)}$	1.00	1.12	1.00	1.68	1.70	2.00

1 = Basal measurements; 2 = sleep in the cold chamber; 3 = cold chamber + exercises.
Differences ($p < 0.05$): *only in Moscow (1–3); **with Moscow (1–1, 2–2, 3–3).

Table 11.9. Changes in the phospholipid spectrum of the blood sera (mg/100 ml) during climatic chamber simulations of the expedition before (Moscow) and after (Ottawa) the ski trek

Phospholipid	Moscow			Ottawa		
	1	2	3	1	2	3
PHH	82.3 ± 2.8	89.2 ± 3.2	83.0 ± 1.8	83.7 ± 1.2	80.7 ± 1.6	80.1 ± 2.5
PHEA	55.0 ± 1.4	58.5 ± 2.5	62.1 ± 1.8	62.4 ± 1.5	59.9 ± 1.6	57.8 ± 2.5
PHS	31.3 ± 3.5	30.9 ± 2.7	33.8 ± 1.7	32.7 ± 2.3	34.3 ± 2.4	30.5 ± 1.5
CL	10.2 ± 1.8	8.5 ± 0.9	12.8 ± 1.9	15.9 ± 1.1	18.4 ± 1.3	15.3 ± 1.8
S	26.2 ± 1.4	24.9 ± 0.5	24.1 ± 1.6	21.1 ± 1.4	17.0 ± 1.4	18.6 ± 1.1
LF	3.7 ± 1.3	5.7 ± 0.4	6.8 ± 0.4	4.6 ± 0.5	6.1 ± 0.4	6.4 ± 0.6
Total	208.7 ± 7.9	217.7 ± 8.1	222.6 ± 4.7	220.2 ± 7.3	216.3 ± 4.9	208.7 ± 7.2

PHH = Phosphatidycholine; PHEA = phosphatidylethanolamine; PHS = phosphatidylserine; CL = cardiolipin; S = sphingomyelin:, LF = lysoforms.
1 = Basal measurements 2 = sleep in the cold chamber; 3 = cold chamber + exercises.

Table 11.10. Changes in the apoprotein concentrations (mg/dl) of the blood sera during the ski trek

Apoprotein	Moscow	Dikson	Stage I	Stage II	Stage III	Ottawa
Apo-A_1	131 ± 5 (n = 8)	143 ± 4 (n = 13)	163 ± 6* (n = 10)	164 ± 8* (n = 7)	166 ± 4* (n = 10)	152 ± 7* (n = 11)
Apo-B	95 ± 7 (n = 8)	96 ± 6 (n = 13)	87 ± 5 (n = 10)	80 ± 8 (n = 7)	81 ± 4 (n = 7)	84 ± 4 (n = 13)

*Difference ($p < 0.05$) from initial (Moscow) data.

Table 11.11 Changes in the apoprotein concentrations (mg/dl) during the climatic chamber simulations of the expedition before (Moscow) and after (Ottawa) the ski trek

Apoprotein	Moscow			Ottawa		
	1	2	3	1	2	3
Apo-A_1	131 ± 5 (n = 8)	118 ± 5 (n = 6)	141 ± 7 (n = 10)	152 ± 7* (n = 11)	160 ± 6* (n = 9)	161 ± 4* (n = 7)
Apo-B	95 ± 7 (n = 8)	91 ± 9 (n = 6)	88 ± 5 (n = 8)	84 ± 4 (n = 13)	88 ± 6 (n = 7)	96 ± 6 (n = 9)

1 = Basal measurements; 2 = sleep in the cold chamber; 3 = cold chamber + exercises.
*Differences ($p < 0.05$) with Moscow data (1-1, 2-2, 3-3).

Serum Lipid Profile during Chamber Simulations of Trek

Blood levels of total lipids and (VLDL + LDL) were higher in Moscow than in Ottawa, presumably reflecting lower rates of blood utilization of fat before the expedition had begun. Blood concentrations of triglyceride, and phospholipids remained unchanged during chamber simulations of the trek (table 11.7–11.9).

Neither sleeping in the cold chamber nor the physical activity decreased total lipids in Moscow, but there was some decrease in total lipids in Ottawa, more marked after sleep than after the subsequent physical activity (table 11.7).

Apoprotein Concentrations during the Expedition

The apoprotein Apo-A_1 was increased at all stages of the traverse (table 11.10). These data together with the low concentrations of VLDL + LDL and the increased levels of HDL show an enhanced metabolism of LDLs, with conversion of VLDL into HDL particles.

Apoprotein Concentrations during Chamber Simulations of the Trek

No changes in apoproteins were observed during chamber simulations of the expedition (table 11.11).

Phospholipid Spectrum during the Trek

There were only small changes in the total pool of phospholipids over the course of the expedition (table 11.6), with low values after 2 weeks of skiing and then a gradual recovery to somewhat above the initial values when tested in Ottawa.

Fractions that decreased included phosphatidylserine, sphingomyelin and particularly phosphatidylcholine. In contrast, the levels of phosphatidylethanolamine and cardiolipin increased in the latter half of the trek. Lysoforms were greatest on Dikson Island and at the North Pole; these compounds are linked to an increased activity of LCAT, and are also related to the appearance of lysosomal hydrolases such as phospholipase-2 in the blood stream.

It is difficult to offer a comprehensive explanation for the changes in phospholipid profile. The overall decrease in phospholipids is probably related to the increased metabolic usage of fat, and this could also explain the decrease in some individual constituents. The increased levels of phosphatidylethanolamine and cardiolipin are conceivably related to the shift in the overall lipid profile to an increased concentration of HDL (since these particles contain more phospholipids than VLDL and LDL).

Phospholipids during Chamber Simulation of the Trek

The chamber experience did not change total blood phospholipids significantly (table 11.9), but substantially increased lysoform concentrations were seen during simulations of the trek, while sphingomyelin concentrations decreased. In Moscow, but not in Ottawa, there was also a trend to an increase in phosphatidylethanolamine concentrations during the chamber experiments.

Conclusions

All the changes in lipid profile point to an increased usage of fat when subjects were exposed to a combination of cold and heavy physical work. The most obvious changes are in the transporting lipoproteins, with a shift towards HDL, increased apo-A and decreased apo-B. Plainly, there is a rapid metabolic degradation of VLDL under the influence of LPL, LCAT, heparin-independent triglyceride lipase and other blood hydrolases, with increased rates of LDL metabolism in the liver. However, blood levels of FFAs did not rise, because active liberation of fatty acids was counterbalanced by rapid fat metabolism in muscle and other tissues. The decrease in phosphatidylcholine and the increase in lysophosphatides suggests that the phospholipids also played a significant role in energy metabolism and the transformation of VLDL into HDL.

Lipid Peroxidation

General Considerations

A slow peroxidation of lipids occurs in living tissues under the action of free radicals, this process being accelerated by exercise, ozone exposure and other extreme conditions. The peroxidation is harmful to both cell membranes and enzyme function. Enzymes of the glycolytic pathway, Krebs cycle and the mithochondrial respiratory chain seem particularly susceptible to peroxidation. The peroxides inhibit cell division, and by damaging the membranes of lysosomes, they can initiate cytolysis. Free radicals attack the most important structural components of the cell membrane, including phosphatidylcholine, phosphatidylethanolamine and cardiolipin. However, the protein fractions usually show no increase in free radicals. Typically, formation of hydroperoxides and alkylperoxides is associated with the presence of polyunsaturated fatty acids in the phospholipids. Phosphatidylethanolamine, for example, contains more polyenoic fatty acids than does phosphatidylcholine, and is thus more vulnerable to free radicals.

In an aqueous medium, interactions between proteins and fatty acids are hydrophobic in nature. Lipid peroxidation disturbs this hydrophobic process, thereby changing the orientation of the protein molecules relative to the surface of the membrane, and disturbing their function. The loosening of lipid–lipid and protein–lipid interactions may lead to the appearance of hydrophilic pores in the cell membrane. For example, erythrocytes undergo hemolysis when they accumulate a finite total of hydroperoxide molecules. Likewise, an accumulation of a specific number of hydroperoxides in a given lysosome causes an active release of lysosomal enzymes into the surrounding medium.

Lipid peroxidation may be associated with the activity of enzymes such as catalases and peroxidases, but it can also be catalyzed nonenzymically through the action of metals with a transient valency (for example, iron or copper). Heme iron is much more active than non-heme iron in this regard. The inhibitors of the hemoproteins are compounds such as tocopherols, glutathione, $NADPH^+$ and ascorbic acid. α-Tocopherol, or vitamin E, is the most potent protector of the red cell membrane. It can reduce Fe^{3+} to Fe^{2+}, and it can also prevent the formation of lipid–protein covalent bindings. In the absence of vitamin E, the fluorescent products of lipid peroxidation are easily formed in the red cell membrane, due to interactions of short-chained dialdehydes with aminophospholipids. The tocopherol is transported in the plasma in association with LDLs.

Tocopherol Determinations

Plasma and erythrocyte tocopherol levels were determined by the method of Taylor et al. [1976], with preliminary removal of saponified fat fractions in the presence of ascorbic acid. Tocopherol, transferred into the hexan phase, was measured fluorimetrically at wavelengths of 286 and 334 nm, absorbance being compared with Sigma standard preparations of tocopherol.

Peroxidation during the Expedition

Plasma tocopherol was significantly increased when the skiers reached Dikson Island (table 11.12). After 2 weeks of skiing, levels had decreased relative to this peak; at 4 weeks, levels were normal, but they were again reduced at the North Pole. These observations suggest that there may have been some deficiency of antioxidant in the diet and/or an increase in peroxidation during the trek. Blood levels of malone dialdehyde, a product of lipid peroxidation, were similar in Moscow and at Dikson. During the traverse, malone dialdehyde concentrations increased, but there was a return to normal values in Ottawa, soon after the trek was completed.

Studies of erythrocytes demonstrated parallel changes (table 11.12), with increased readings at Dikson, but a drop over the trek which persisted into the final observations in Ottawa. The fluorescent products of lipid

Table 11.12. Changes in tocopherol and the products of lipid peroxidation (PLP) in plasma and erythrocytes during the ski trek

	Moscow	Dikson	Stage I	Stage II	Stage III	Ottawa
Tocopherol in plasma mg/100 ml	1.03 ± 0.10 (n = 7)	1.78 ± 0.14* (n = 13)	0.82 ± 0.17* (n = 7)	1.07 ± 0.14 (n = 8)	0.67 ± 0.10** (n = 7)	1.06 ± 0.03 (n = 11)
Malondialdehyde in plasma, nmol/ml	2.7 ± 0.11 (n = 5)	2.6 ± 0.12 (n = 13)	3.9 ± 0.12** (n = 7)	4.26 ± 0.21* (n = 8)	4.0 ± 0.42* (n = 9)	2.4 ± 0.18 (n = 11)
Tocopherol in erythrocytes, mg/100 ml	0.3 ± 0.02 (n = 7)	0.48 ± 0.01* (n = 13)	0.22 ± 0.03** (n = 7)	0.23 ± 0.04* (n = 7)	0.19 ± 0.02** (n = 8)	0.15 ± 0.06** (n = 13)
PLP in erythrocytes, relative units	0.88 ± 0.08 (n = 8)	1.07 ± 0.36 (n = 13)	2.25 ± 0.21** (n = 2)	1.30 ± 0.12* (n = 6)	0.91 ± 0.04 (n = 6)	1.04 ± 0.11 (n = 12)

Differences ($p < 0.05$): *from Moscow data; **from Dikson data.

peroxidation in the erythrocytes increased against this background of antioxidant deficiency, reaching a maximum after 2 weeks of skiing, with a subsequent gradual return towards normal values.

Analyses of the phospholipid spectra for the erythrocyte membranes further demonstrated an increased content of cardiolipin and lysoforms. However, there was a compensatory increase in superoxide dismutase in the erythrocytes, from the norm of 84.0 ± 7.5 units to a value of 190 ± 37 units in Ottawa, after completion of the expedition. One factor contributing to this apparent compensation for oxidant activity may have been the appearance of young erythrocytes in the peripheral blood.

Although the changes in lipid peroxidation were observed for convenience in the erythrocytes, it is probable that similar changes in function occurred in the liver and other metabolically active tissues.

Changes in Erythrocyte Glycolysis during the Expedition

General Considerations

The erythrocyte provides a convenient model of probable overall changes in cell metabolism. However, when subjects are exercising hard, there may also be changes designed to modify the oxygen-transporting ability of the red cells.

Glycolysis is the sole source of energy for adenosine triphosphate (ATP) synthesis in the erythrocytes. It is also important to the diphosphoglycerate shuttle concerned in the synthesis of 2,3-diphosphoglycerate, this mechanism accounting for about 20% of the overall erythrocyte glucose metabolism. An adult man has about 5 mg 2,3-diphosphoglycerate/ml cells, with a half-life of some 7 h. This substance strongly diminishes the affinity of haemogloblin for oxygen, thus shifting the oxygen dissociation curve to the right. Oxygen affinity is also influenced by pH, an enhanced lactate production in the erythrocytes decreasing pH and shifting the dissociation curve to the right. A decrease in erythrocyte glycolysis can thus increase oxygen affinity, helping oxygen uptake in the lungs, but making it more difficult to extract oxygen in the tissues.

Four key enzymes in the glycolytic pathway catalyze thermodynamically irreversible reactions: hexokinase, the primary phosphorylation of glucose; phosphofructokinase, additional phosphorylation of glucose-6-phosphate; phosphoglycerate kinase and pyruvate kinase. The first two reactions consume ATP, while the third enzyme catalyses the synthesis of ATP.

Hexokinase is easily inhibited by the glucose-6-phosphate that it produces, although the constant of inhibition is decreased by rising local concentrations of ATP and inorganic phosphate. Glucose-1,6-diphosphate is even more effective in this regard. Under conditions of stress, the activity of hexokinase is modulated by hormones, glucocorticoids inhibiting and insulin serving as an activator of the enzyme.

Phosphofructokinase has a very complex allosteric regulation. It is activated by its substrate (fructose-6-phosphate), its product (fructose-1,6-diphosphate), and also by adenosine diphosphate, adenosine monophosphate, and inorganic phosphate. ATP has a dual action. As the coenzyme of phosphofructokinase, it activates the enzyme but, as an allosteric effector, it hampers the reaction. Activators of phosphofructokinase operate by diminishing the sensitivity of the enzyme to the inhibitory action of ATP. Typically, stress diminishes the activity of phosphofructokinase.

Phosphoglycerate kinase catalyzes the formation of ATP. Regulation of this step in glycolysis is closely bound to the activity of triosephosphate dehydrogenase. NADH and 1,3-diphosphate inhibit the enzyme in accordance with the principles of negative feedback. Phosphoglycerate kinase is specifically activated by the mitochondrial protein kinasine; this mechanism is absent in erythrocytes, since they contain no mitochondria.

Pyruvate kinase catalyses the glycolytic formation of ATP through the breakdown of phosphoenolpyruvate. The enzyme is strongly inhibited by ATP, and is also regulated by the phorphorylation-dephosphorylation mechanism. Both pyruvate kinase and phosphoglycerate kinase are inhibited by diphosphoglycerate. Hexokinase and phosphofructokinase are less sensitive to diphosphoglycerate.

Methodology

The glycolysis rate of the erythrocytes was determined in a 'reconstructed' medium with optimum concentrations of activator and cofactors [Tretyakova and Panin, 1978]. The blood was defibrinated, centrifuged, washed, hemolyzed with distilled water, and then centrifuged again for 1 h to precipitate the stroma, before resuspending in the optimized buffer. Different substrates were then added, for example glucose, glucose-6-phosphate, and fructose-1,6-phosphate. Glycolytic activity was expressed in terms of lactate formation.

Changes in Erythrocyte Glycolysis over the Expedition

Use of the three substrates glucose, glucose-6-phosphate and fructose-1,6-diphosphate allowed an assessment of limitations of glycolysis at the hexokinase, phosphofructokinase and probably the pyruvate kinase levels. Typical initial rates of glycolysis (per liter of erythrocytes per hour) were 4.0–4.5 mmol glucose, 19–23 mmol glucose-6-phosphate and 25–26 mmol fructose-1,6-diphosphate (table 11.13). These values lie at the bottom of the normal range. The limiting step was hexokinase, since the rate of glycolysis was enhanced if glucose-6-phosphate was provided, with phosphofructokinase as the next limiting step.

In Dikson, there were insignificant trends to increases in the metabolism of glucose and glucose-6-phosphate. Normal rates of glycolysis were seen over the trek, but final values in Ottawa were reduced, this data coinciding with the lower insulin readings obtained in Moscow and Ottawa. In the early stages of the trek, metabolism of glucose-6-phosphate was slowed, with later recovery, while the metabolism of fructose-6-phosphate remained low throughout the trek and after the mission had been completed. This suggests that during the traverse, regulation of glycolysis was shifted from hexokinase to a lower point in the metabolic chain; this could arise in part through the various allosteric regulating mechanisms discussed above, and in part through the increased levels of glucocorticoids. This response can be explained in terms of the need to transport substantial quantities of oxygen during the traverse. Glucose-6-phosphate can be metabolized either by glycolysis or by the hexose monophosphate shunt, or it can be converted to glucose. If phosphofructokinase is inhibited, glucose is formed and directed through the pentose phosphate pathway. This keeps diphosphoglycerate at a low level and assures a high affinity of hemoglobin for oxygen. Other studies of expeditions in the Arctic and Antarctica have reached similar conclusions.

Table 11.13. Changes in the rate of erythrocyte glycolysis (mmol lactate/liter erythrocytes/h) during the ski trek

Substrate	Moscow	Dikson	Stage I	Stage II	Stage III	Ottawa
Glucose	3.90 ± 0.26 (n = 7)	4.70 ± 0.35 (n = 13)	4.50 ± 0.30 (n = 8)	4.50 ± 0.41 (n = 7)	4.90 ± 0.32* (n = 10)	3.10 ± 0.10** (n = 13)
Glucose-6-phosphate	19.6 ± 1.2 (n = 7)	25.0 ± 1.4* (n = 13)	14.8 ± 1.0** (n = 8)	13.8 ± 1.4** (n = 7)	18.7 ± 1.2* (n = 10)	18.7 ± 0.7* (n = 13)
Fructose-1,6-diphosphate	26.0 ± 1.0 (n = 7)	25.0 ± 1.1 (n = 13)	19.2 ± 0.9** (n = 8)	18.8 ± 0.9** (n = 7)	18.4 ± 0.8** (n = 10)	17.1 ± 0.5** (n = 13)

Differences ($p < 0.05$): *from Moscow data; **from Dikson data.

Table 11.14. Changes in the rate of erythrocyte glycolysis (mmol lactate/liter/erythrocytes/h) during climatic chamber simulations of the expedition before (Moscow) and after (Ottawa) the ski trek

Substrate	Moscow			Ottawa		
	1	2	3	1	2	3
Glucose	3.91 ± 0.26 (n = 7)	3.90 ± 0.60 (n = 8)	4.40 ± 0.20 (n = 8)	3.10 ± 0.10* (n = 13)	3.00 ± 0.36 (n = 5)	3.20 ± 0.10* (n = 6)
Glucose-6-phosphate	19.6 ± 1.2 (n = 7)	19.4 ± 1.3 (n = 8)	20.8 ± 1.2 (n = 8)	18.7 ± 0.7 (n = 13)	20.3 ± 1.4 (n = 5)	20.0 ± 2.0 (n = 6)
Fructose-1,6-diphosphate	26.0 ± 1.0 (n = 7)	23.8 ± 1.7 (n = 8)	24.9 ± 2.3 (n = 8)	17.1 ± 0.5* (n = 13)	18.7 ± 0.4* (n = 5)	17.6 ± 1.1* (n = 6)

1 = Basal Measurements; 2 = sleep in the cold chamber; 3 = cold chamber + exercise.
*Differences ($p < 0.05$) with Moscow data (1–1, 2–2, 3–3).

It should be stressed that the erythrocyte-based study of glycolytic regulation cannot necessarily be translated directly into a whole body response. Hormonal regulation of glycolysis is likely to be particularly marked in tissues like liver and muscle that have well-developed receptors for glucocorticoids, insulin and glucagon.

Climatic Chamber Simulations of Expedition

In Moscow, neither sleeping in the cold nor the acute exercise stress displaced the limiting step in the glycolytic reaction (table 11.14). In the Ottawa chamber exposure, after completion of the trek, the absolute rates were lower for both glucose and fructose-1,6-diphosphate metabolism, but again there was no shift in the limiting step in the enzyme chain. These findings may be interpreted as evidence for a combination of hormonal and allosteric downregulation of glycolysis.

Conclusions

The inhibition of glycolysis, seen here, serves to supplement the enhanced lipolysis discussed in the previous section. Other evidence of the shift to lipid metabolism can be found in determinations of triglyceride lipase and lipoprotein lipase activity, comparisons of arteriovenous differences for glucose and for FFAs, and a greater shift towards HDL in venous than in capillary blood.

The hormonal basis of the shift to fat metabolism seems an increase in glucocorticoids and glucagon while insulin levels remain unchanged. Although this represents a moderate metabolic strain, in general there remained some potential for further adjustment through a decreased production of insulin; however, this last mechanism was adopted during the cold-chamber experiments in Ottawa, giving rise to what the Soviet authors designate as a 'strain diabetes'. There was an associated increase in protein metabolism over the trek, as shown by the development of hypoalbuminemia.

The increase in fat and protein metabolism induced structural, age-like changes in cell membranes, and these changes were enhanced by lipid peroxidation and the release of lysozymes. While there is normally little evidence of tocopherol deficiency in the diet, under such extremes of stress, there is plainly a greater possibility that improved nutrition could play a compensatory role.

Acid Lysosomal Hydrolases

General Considerations

Lysosomes are present in almost all cells. They contain acid hydrolases which can partly or completely split most compounds that contain simple or complex ethers. Normally, these potent enzymes lie dormant within the lysosome, but when an organism is exposed to extreme conditions, the lysosomes increase rapidly in number and in size, and the permeability of their membranes is also increased.

The structural and functional changes in the lysosomes can be related to the severity of the functional strain, as measured by catecholamines and glucocorticoid secretion. Initially, these changes are adaptive rather than harmful, but if the strain is excessive, the structural integrity of the lysosome membrane is damaged, and the acid hydrolases escape into the cytosol, damaging the tissue.

Pharmacological doses of glucocorticoids have a stabilizing influence upon lysosome membranes. However, glucocorticoids can also facilitate proteolysis through their action upon hepatic lysosomal enzymes. Catecholamines increase the tissue activity of lysosomal enzymes, thus enhancing the catabolism of proteins in the liver. Glucagon also has a marked influence upon the lysosomal apparatus, inducing cytolysis. In contrast, insulin inhibits tissue proteolysis.

In animals with experimental diabetes, treatment with glucocorticoids and catecholamines increases the number of secondary lysosomes in the liver, heart and skeletal muscles. The lysosomes thus formed have a high specific hydrolase activity and an increased membrane permeability. Similar changes are induced by exposing the organism to various types of stress.

Phagocytes are a second source of lysozymes in the blood stream, the main source being the monocytic phagocytes. These cells migrate to areas of tissue damage, releasing both lysozymes and other mediators that facilitate healing. Enzymes such as cathepsin-D and acid phosphatase are actively consumed by the proliferating somatic cells as tissue repair occurs.

Methodology

Lysosomal enzymes were determined in terms of the product yielded when incubated with appropriate substrates [Barrett, 1972]. The hydrolytic products from cathepsin-D and acid DNAase were determined by direct spectrophotometry, while color reactions were used to assay the products from β-glucosidase, β-galactosidase and acid phosphatase (n-nitrophenol and inorganic phosphate reagents, respectively).

Changes in Lysozymes Observed during the Expedition

Acid phosphatase levels were increased at Dikson, but remained below the initial Moscow values throughout the trek (table 11.15). Cathepsin-D values were also reduced both during the traverse and subsequently in Ottawa. β-Galactosidase was increased at Dikson relative to the Moscow baseline, and remained increased during the first half of the mission, but later dropped to low levels. Acid DNAase was high throughout the expedition, with particularly high readings at Dikson and in Ottawa. The increased concentrations of these last two enzymes suggest that destructive processes were enhanced in response to the prolonged and severe physical strain of the trek. If so, it is surprising that there were not also increments in cathepsin-D and acid phosphatase. Possibly the latter two enzymes were actively involved in the recuperative processes.

Table 11.15. Activity of blood lysosomal enzymes (mmol substrate/l/min) during the ski trek

Enzymes[1]	Moscow	Dikson	Stage I	Stage II	Stage III	Ottawa
Acid phosphatase	62.2 ± 8.08 (n = 7)	86.7 ± 3.56* (n = 13)	27.2 ± 2.25* (n = 7)	21.9 ± 4.54* (n = 6)	22.0 ± 3.73* (n = 8)	44.6 ± 4.17* (n = 13)
Cathepsin-D	28.4 ± 7.15 (n = 7)	18.4 ± 0.49 (n = 13)	8.8 ± 0.86** (n = 7)	11.9 ± 1.17** (n = 6)	9.6 ± 0.52** (n = 8)	8.6 ± 0.42** (n = 13)
β-Galactosidase	33.3 ± 1.84 (n = 7)	53.4 ± 7.21* (n = 13)	55.7 ± 8.43* (n = 7)	66.8 ± 15.3 (n = 6)	25.7 ± 8.53* (n = 8)	22.8 ± 2.72** (n = 13)
Acid DNA-ase	36.8 ± 6.49 (n = 7)	93.2 ± 20.4* (n = 13)	53.5 ± 7.02 (n = 7)	58.9 ± 12.4 (n = 6)	46.6 ± 3.93* (n = 8)	68.0 ± 12.33* (n = 13)

Differences (p < 0.05): *from Moscow data; **from Dikson data.
[1]Substrates: Acid phosphatase = β-glycerophosphate Na salt; cathepsin-D = hemoglobin; β-galactosidase = n-nitrophenyl-p-D-galactopiranoside; acid DNA-ase = DNA.

Table 11.16. Changes in the activity of blood lysosomal enzymes (mmol substrate/l/min) during climatic chamber simulation of the expedition before (Moscow) and after (Ottawa) the ski trek

Enzyme	Moscow			Ottawa		
	1	2	3	1	2	3
Acid phosphatase	62.2 ± 8.08 (n = 7)	60.8 ± 8.09 (n = 7)	78.1 ± 2.91 (n = 10)	44.6 ± 4.17 (n = 13)	36.9 ± 7.08** (n = 10)	60.9 ± 4.72** (n = 9)
Cathepsin-D	28.4 ± 7.15 (n = 7)	7.7 ± 0.35* (n = 7)	17.5 ± 0.90 (n = 10)	8.6 ± 0.42** (n = 13)	7.3 ± 0.47 (n = 10)	2.5 ± 0.88** (n = 9)
β-Galactosidase	33.3 ± 1.84 (n = 7)	35.2 ± 7.78 (n = 13)	25.1 ± 6.53 (n = 7)	22.8 ± 2.72** (n = 6)	26.7 ± 3.50 (n = 8)	21.5 ± 5.53 (n = 13)
Acid DNA-ase	36.8 ± 6.49 (n = 7)	46.6 ± 13.3 (n = 13)	49.6 ± 3.70 (n = 7)	68.0 ± 12.3** (n = 6)	92.8 ± 17.6 (n = 8)	77.0 ± 13.91 (n = 13)

Differences (p < 0.05): *only in Moscow (1–2); **with Moscow (1–1, 2–2, 3–3).

Changes in Lysozymes during Chamber Simulations of the Expedition
During the initial chamber experiments in Moscow, the only significant change in the lysozymes was a decrease in cathepsin-D during sleep, this mirroring the decrease seen during the trek (table 11.16). In Ottawa, after completion of the expedition, the responses seen during the trek were followed even more closely; there was a decrease in cathepsin-D, a biphasic change in acid phosphatase, and a tendency to increments of β-galactosidase and acid DNAase.

In all, the changes in lysozymes are consistent with a moderate degree of strain during both the trek and its chamber simulation.

Changes in Structure and Function of Erythrocyte Membranes

Methodology

Erythrocyte 'shadows' were obtained at Dikson, the North Pole, and in Ottawa. The objectives of this investigation were to study the viscosity, specific electrical conductivity, and the activation energy associated with ionic transport, to analyze the structure of the cell membrane by infrared spectroscopy, and to determine K_m and V_{max} for the Na^+K^+-ATPase. Observations were made over the temperature range 34–42 °C.

Viscosity and electrical conductivity were plotted against temperature, and the points of phase transition (Tc) were determined for each sample. A shift in phase transition reflects structural and functional changes in the membrane, with implications for Na^+K^+ATPase. The activity of this membrane-bound enzyme depends strongly upon the degree of order in the membrane (that is, upon the content of polyenoic fatty acids and cholesterol, the type of phospholipids, and the presence of their lysoforms).

The viscosity was measured in an ultrathermostat, by traditional capillary viscosimetry. The speed of flow of a suspension of erythrocyte shadows through a capillary tube of known diameter (0.34 mm) and length (100 mm) was determined under a standard driving pressure (20 mm H_2O). Viscosity was measured at intervals of 1 °C over the 34–42 °C range, calculations being based on the equation:

$$P_x = P_o \frac{t_x}{t_o}$$

where P_o and P_x are the viscosities of the phosphate buffer and the suspension of erythrocyte shadows respectively, and t_x and t_o are the times for a fixed volume of the fluid (0.1 ml) to pass through the tube. Over the chosen temperature range, the viscosity of the buffer was 1 cP. The viscosity of the erythrocyte shadows was estimated to within 1% relative to this standard.

The electrical conductivity of a colloidal solution depends on the disperse and the dispersive phases of the system (the number and motility of the colloidal particles, the size of their charge, and the number and charge of the ions in solution.

The charge on the particles is given by:

$$Q = z \cdot e \cdot k(1 + x \cdot z)$$

where z is the size of the particle, e is the dielectric strength, k is the electrokinetic potential, and x is the thickness of the double electrical layer around a colloidal particle. The specific conductivity C is then given by:

$$C = Q \cdot v \cdot U_o$$

where v is the number of particles in 1 cm³, as determined from the ratio:

$$v = \frac{0.01}{\frac{4}{3}\pi z^3}.$$

The term 0.01 assumes that 1% of colloidal particles are in the solution. U_o is the cataphoretic mobility, determined from the expression:

$$U_o = \frac{e \cdot k}{6\pi P}$$

where P is the viscosity of the solution. By substitution,

$$C = \frac{0.01 \, e^2 \cdot k^2(1+x^2)}{8\pi^2 z^2 P}.$$

From this final equation, it emerges that the electrical conductivity of a colloidal system is strongly influenced by the particle's electrokinetic potential, the dielectric strength, the particle size, the thickness of the double layer and the viscosity of the solution. The equations presented are strictly limited to particles around 50 Å in size, but the concepts can be extended to larger particles, provided that the concentration and surface potential of the colloid, and the concentrations of the potential-forming ions in solution are all low.

Biological membranes display some of the properties of liquid crystals. The temperature dependence of conductivity of the erythrocyte shadows should thus be given by the expression:

$$C = C_o \, e^{\frac{-Ea}{KT}}$$

where C_o is a constant depending on the nature of the particle, K is Boltzmann's constant, Ea is the activation energy, and T is the absolute temperature. The temperature dependence should also be related to the number and motility of the carriers of current, according to the equation:

$$C = n \cdot e \cdot m_o$$

where n is the number of free charges, m_o is the motility of the charge, and e is the charge on an electron.

The experimental technique measured the resistance of suspensions of erythrocyte shadows when these were placed in a cuvette with electrodes of known surface, separated by a known distance. A 6-V, 2-kHz voltage was applied. Conductivity was then obtained as the inverse of specific resistance. The concentration of the shadows was 0.46 mg/ml, measured as a protein concentration, while the dispersing sodium/potassium/phosphate buffer had a pH of 7.4 and an osmolality of 310 osm.

Structural changes in the erythrocyte membrane were also evaluated by infrared spectroscopy. A suspension of the erythrocyte shadows (0.2 ml) was placed in a cuvette with a calcium fluoride base, and was vacuum dried for 1.0–1.5 h at 3–4 °C. The film of dried erythrocytes was then treated twice with phosphate buffer and was incubated at the desired temperature for 30 min before vacuum drying at 21–23 °C for a further 15 min. Spectra were determined on an UR-20 infrared spectrophotometer (GDR). Measurements were made of the half widths of absorption at 1660 cm^{-1} (corresponding to the C=O bonds in proteins), 1,000–1,100 cm^{-1} (the P—O—C bond in phospholipids), 1,500–1,700 cm^{-1} (C=O and NH bonds in proteins), and 2,700–3,000 cm^{-1} (the C—H bonds in proteins and phospholipids), together with the maximum intensity of absorption at 1,745 cm^{-1} (C=O bond in phospholipids) and the background intensity at 1,480 cm^{-1}.

Since the erythrocyte membrane contains many proteins, it was not possible to translate the spectral absorption into standard units of liters per mole per centimeter. Rather, the integral of the absorption peak was expressed as square centimeters per gram of erythrocyte shadow. An essentially equal total mass of erythrocytes was used for preparing test films at the three sites (Dikson, 0.50 ± 0.05, North Pole 0.49 ± 0.09, and Ottawa 0.41 ± 0.06 mg).

The spectrophotometer was calibrated using a polystyrol film. The relative error of the absorption measurements was estimated at 1%.

Changes in Viscosity during the Expedition

The initial viscosity of the erythrocyte shadow suspensions at a temperature of 34 °C was 1.85 cP. The viscosity decreased rather uniformly as temperature increased, both in Dikson and in Ottawa (fig. 11.1), but the curve had an anomalous region between 36 and 38 °C.

In Canada, the phase transition was displaced upwards, by 0.5 °C, and changes in viscosity of the erythrocyte shadows were more marked than in Dikson or at the North Pole (table 11.17). The data thus suggest that an increase in rigidity of the erythrocyte membranes developed during the

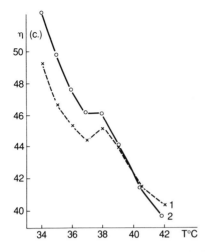

Fig. 11.1. Temperature dependence of viscosity of erythrocyte shadows on Dikson Island (1) and in Ottawa (2).

Table 11.17. Changes in viscosity of erythrocyte 'shadows' over the temperature range 34–42 °C (mean ± SD)

Parameters	Origin of the blood sample		
	Dikson Island	North Pole	Canada
Point of phase transition, °C	36.7 ± 1.2	36.7 ± 1.4	37.2 ± 1.1
Viscosity changes, cP	0.305 ± 0.063	0.220 ± 0.091	0.439 ± 0.136

course of the trek, which reversed after completion of the mission. In the view of the Soviet investigators, these changes can probably be attributed to increases in lysoforms, phospholipids, and lipid hydroperoxides. Not only would such changes disturb hydrophobic interactions, making the membranes more porous, but they would presumably increase interactions between proteins and lipids, making the membrane more stable. Later, these changes may have been offset by an increase in the number of newly formed erythrocytes, giving the increased delta viscosity seen in Ottawa.

Changes in Electrical Conductivity during the Expedition

At low temperatures, conductivity showed an almost exponential relationship to temperature, but at high temperatures the relationship

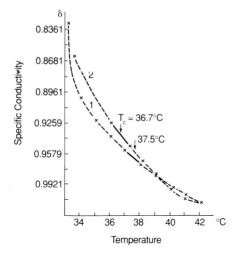

Fig. 11.2. Temperature dependence of electrical conductivity of erythrocyte shadows on Dikson Island (1) and in Ottawa (2).

Table 11.18. Changes in the phase transition point and intrinsic electric conductivity of erythrocyte 'shadows' over the temperature range 34–42 °C (mean ± SD)

Parameters	Origin of the blood sample		
	Dikson Island	North Pole	Canada
Phase transition point, °C	37.2 ± 1.2	37.6 ± 1.0	37.8 ± 1.0
Changes in intrinsic electrical conductivity $\Omega^{-1} \times m^{-1}$	0.131 ± 0.035	0.101 ± 0.029	0.105 ± 0.022

became more linear (fig. 11.2). The transition from one pattern of temperature dependence to the other was smoother than that observed in homogenous liquid crystals.

The transition points showed no significant differences between the blood samples collected on Dikson Island, at the North Pole and in Ottawa (table 11.18). However, differences in activation energy before and after the phase transition point indicate the structural change in the cell membrane at 37–38 °C. Moreover, the ratios of activation energies at low and high temperatures changed over the trek, being highest at the North Pole (table 11.19). The increase in activation energy at the North Pole, and subse-

Table 11.19. Activation energy (EA) of erythrocyte 'shadows' at low temperature (Ea_1) and high temperature (Ea_2) parts of the dependence curve (mean $\pm$ SD)

Origin of the sample	Activation energy, cal/mg		
	Ea_1	Ea_2	Ea_1/Ea_2
Dikson Island	6,492 $\pm$ 2,487	5,106 $\pm$ 2,955	1.643 $\pm$ 0.475
North Pole	7,325 $\pm$ 2,707	3,148 $\pm$ 1,085	2.306 $\pm$ 1.051
Canada	11,771 $\pm$ 2,494	5,592 $\pm$ 2,045	2.108 $\pm$ 1.012

Table 11.20. Temperature depedence of integrated intensity of the infrared spectra of erythrocyte 'shadows' (mean $\pm$ SD)

Origin of the sample	Absorption intensity, %ΔI between 40 and 37 °C		
	1,660 cm^{-1}	1,070 cm^{-1}	I1,660 cm^{-1}/I1,070 cm^{-1}
Dikson Island	8.87 $\pm$ 3.48	9.85 $\pm$ 3.67	1.82 $\pm$ 0.37
North Pole	7.52 $\pm$ 4.00	6.92 $\pm$ 2.31	2.61 $\pm$ 0.21
Canada	5.87 $\pm$ 3.12	7.91 $\pm$ 4.75	2.54 $\pm$ 0.36

quently in Ottawa, suggests that the small ions (H^+, K^+, Na^+ etc) had less ability to compensate for their particulate charge in a suspension of membranes with a lesser structural integrity.

Changes in Infrared Absorbance during the Expedition

Absorption in the regions of C=O bond (1,660 cm^{-1}) and the NH bond (1,550 cm^{-1}) decreased significantly as the temperature of the measurements was reduced (fig. 11.3–11.6), these changes being more marked in the initial blood samples from Dikson Island than in the final specimens taken in Ottawa (table 11.20). The Dikson specimens also showed a significant decrease in absorption at 1,645–1,650 cm^{-1}, and at 1,635, 1,685 and 1,696 cm^{-1} (fig. 11.3). The difference between the two locations again points to the development of structural changes in the proteins of the erythrocyte membrane.

As the measuring temperature was decreased, there was also a change in absorption at 1,070 cm^{-1}, reflecting changes in the P—O—C grouping of the phospholipids (fig. 11.4–11.7), the difference amounting to 9.8% at Dikson, 6.9% at the North Pole, and 7.9% in Canada (table 11.20). The P—O—C region showed two maxima, at 1,070 and 1,090 cm^{-1}, this being more marked in Dikson than in Ottawa. Thus, changes in phospholipids also seem to have developed over the course of the trek.

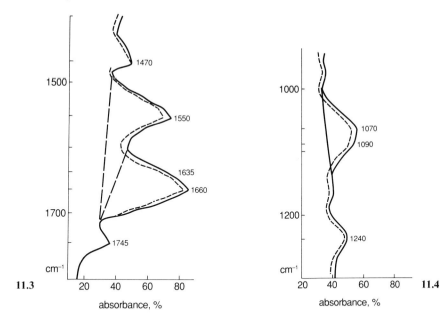

Fig. 11.3. Infrared absorbance of erythrocyte shadows at Dikson Island before trek. —— = 40 °C; - - - - = 37 °C.

Fig. 11.4. Infrared absorbance of erythrocyte shadows at Dikson, before trek. —— = 40 °C; - - - - = 37 °C.

Disarrangement of the structural proteins was further suggested by an increase in absorption at the wavelength of the C=O bond (1,660 cm^{-1}, table 11.21). The decrease in absorption at 1,745 cm^{-1} points to a reduced phospholipid content of the membrane at the North Pole and in Canada, relative to the Dikson baseline (table 11.21).

The intensity of background absorbance at 1,480 cm^{-1} depends on the water content of the sample. Given the standard conditions of vacuum drying, any alteration of this value must reflect changes in the amount of bound water. Values were 10% higher at the North Pole and in Canada, relative to initial blood samples. This suggests that there was an increase in the hydration of the proteins and/or phospholipids.

Integrated absorbance over the range 2,700–3,000 cm^{-1} was increased at the North Pole and in Canada relative to the Dikson baseline (fig. 11.5–11.7). This reflects increased covalent CH bonding of fat and protein. There was an increase in splitting in the lipid parts of the spectrum (particularly at 1,470 cm^{-1}), suggesting deformational oscillation of the CH bonds.

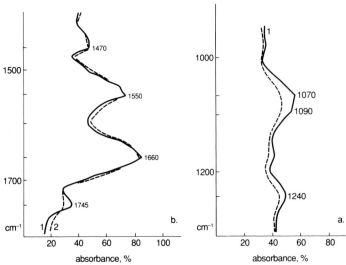

Fig. 11.5. Infrared absorbance of erythrocyte shadows sampled at North Pole. —— = 40 °C; – – – = 37 °C.

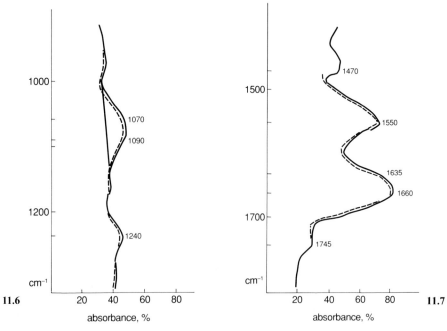

Fig. 11.6. Infrared absorbance of erythrocyte shadows sampled in Canada after trek. —— = 40 °C; – – – = 37 °C.

Fig. 11.7. Infrared absorbance of erythrocyte shadows sampled in Canada after trek. —— = 40 °C; – – – = 37 °C.

Table 11.21. Changes in some characteristics of the infrared spectra of the erythrocyte 'shadows' during the ski trek (mean ± SD)

Origin of the sample	Absorption intensity, %		
	$1,745 \text{ cm}^{-1}$	$1,480 \text{ cm}^{-1}$	$1,660 \text{ cm}^{-1}$
Dikson Island	18.9 ± 5.0	33.1 ± 8.8	64.2 ± 8.6
North Pole	15.5 ± 4.5	41.8 ± 8.0	82.4 ± 10.3
Canada	13.2 ± 3.8	40.9 ± 8.8	80.7 ± 9.5

Changes in ATPase over the Expedition

The activity of ATPase and its K_m for ATP were both decreased in Ottawa relative to initial values in Dikson; from 3.50 ± 0.25 to 2.65 ± 0.32 mmol/mg/h, and from 5.64 ± 0.21 to 3.83 ± 0.72 mmol, respectively.

These changes could be related to a decreased content or an altered composition of phospholipids, together with conformational changes in the enzyme molecule itself.

Conclusions

The changes in hydrophobic interactions and membrane permeability indicated by these various tests are not necessarily negative adjustments, since they enhance the rheological properties of the erythrocytes, and may also enhance the diffusion of oxygen through the cell membranes.

12 Conclusions
Roy J. Shephard

The immediate impression formed from a review of the data collected during the transpolar ski trek is of the tremendous capacity of the human adult to acclimatize both physiologically and psychologically to a very unfavorable environment. Some of the participants in the present expedition were relatively old (a maximum of 51 years), and three of the group showed a substantial resting hypertension. Nevertheless, the entire party were able to complete the 1,800-km traverse with no more permanent adverse effects than two instances of peripheral frostbite.

Given the scientific intent of the expedition, it is somewhat unfortunate that it was not possible to specify the cumulative stress upon the participants more closely. It is known that external temperatures were extremely low over the first month of the trek, but given that the participants were working hard for most of the day, and that they were also well clothed, it may be that the general exposure to cold stress was less than the environmental readings might suggest. Such a speculation is supported by the limited nature of general adaptations to cold. Some observations on British expeditions to Antarctica have even suggested that the problem of polar explorers can be hyperthermia rather than hypothermia. Nevertheless, there was substantial cooling of the extremities, as shown by the amount of frostbite, and the development of local cold acclimatization in the vasculature of the hands. In future expeditions, this issue should be monitored more closely, possibly using the techniques adopted by the British in Antarctica to obtain integrated core and skin temperatures under typical day and night conditions.

The decrease in skinfold readings, loss of body fat and increase in lean tissue mass all echo findings of O'Hara et al. [1978] on Canadian military expeditions to the arctic. The implication seems a high rate of daily energy expenditure, with an overall deficit of energy intake leading to a usage of depot fat. Talk of 'hunger' by the participants and the heart rate estimates of daily energy expenditure support such an explanation, although the calibration of the heart rate recorders while exercising in a temperate environment could overestimate oxygen consumption relative to the polar situation (skiing over rough ice, while exposed to cold and a variety of

12 Conclusions

dangers). The use of double-labelled water or selected measurements of oxygen consumption by means of a portable recorder such as the Oxylog device would provide a more valid estimate of cumulative energy usage on future treks.

A disturbance in fluid balance is suggested by both the large changes in body mass and by the altered blood levels of fluid-regulating hormones. However, more information is needed on the magnitude of this loss, on its distribution between the various body fluid compartments, and its implications for circulatory function and thus aerobic power output. It would be particularly interesting to follow up the possibility that a decrease in blood volume, by its influence upon ventricular preloading, contributed to a decrease in physical working capacity despite a heavy daily work schedule. Perhaps some of the necessary data could be obtained by charting input and output of fluids, and another approach to overall hydration of lean tissue might be measurements of electrical conductivity by battery-operated equipment.

The contribution of the expedition to our more general understanding of the problem of polar stress is less clear-cut. The experiences of a highly motivated and mission-oriented expedition are likely to have only limited application to the behavior of the average citizen who moves to high latitudes. The average citizen faces much less in the way of physical danger, cold exposure, dietary constraints and fluid loss. Moreover, motivations of the average citizen are very different from those involved in a high-profile international expedition, content to live for a long period with almost no external logistic support. If one were to seek practical parallels to the expedition, the closest similarities would be with military expeditions (although these, also, would usually have a stronger logistic support), survival after a plane crash in the arctic (although such individuals would be less well-prepared, both physically and in terms of clothing), and occasional commercial hunting expeditions.

How far was the general problem of stress elucidated? To the extent that acclimatization occurred, the Soviet investigators would argue that there was a 'favorable' dose of stress, to which the body successfully adapted. This concept is supported by various psychophysiological measurements, such as changes in the extent of sinus arrhythmia, an altered variance of neurophysiological reactions, and various biochemical parameters (particularly the relative serum levels of testosterone and cortisol, and the glucagon/insulin ratio). The distinction between 'favorable' and 'unfavorable' amounts of stress is less clear. Future expeditions may well wish to include some measures of immune function, as there is increasing evidence that such data provide important indications of acceptable and excessive levels of stress.

A number of the biochemical indices suggest that the main stress faced during the trek was a high sustained metabolic demand, to which the body responded by switching from carbohydrate to lipid metabolism. The emphasis upon lipid breakdown has practical implications for many of the body lipids, including the serum lipid profile and the structural lipids of the cell membranes. It is uncertain how far the altered viscosity, electrical conductivity and chemical make-up of the cell membranes can be explained by the high rate of lipid utilization, how far it reflects peroxidation of lipids (a consequence of either ozone exposure or a high rate of metabolism), and how far it is merely due to a limited diet. Indeed, it is unclear whether the observed changes, such as an alteration in 2,3-diphosphoglycerate, help or hinder the overall biological response to the new environment. However, it does seem possible that a number of these changes could be reversed by an increased dietary intake of α-tocopherol.

The Soviet investigators were particularly interested in determining who would perform well during the dangers encountered on the trek. Although data from the MMPI and Cattell questionnaires were interpreted in considerable detail in this context, it seems important to stress that these instruments were devised primarily for clinical purposes; there remains a need for simple, well-designed instruments specifically directed to the needs of the normal personality. Given the importance of group interactions to wilderness survival, some measures of abilities at group interaction would be an important facet of such a tool. A further important personal quality, which may influence group acceptance, is the capacity to make a realistic self-appraisal. Finally, the successful participant must be willing to try new ideas, without an excessive fear of failure.

As in the Human Adaptability Project of the International Biological Programme, we may conclude that there is some possibility to match the characteristics of the human adult with the demands of a harsh environment through a combination of selection and preliminary acclimatisation. But all too often anatomical and physiological characteristics that are helpful in meeting one environmental challenge prove a handicap in a second adverse situation. Above all, the challenge is mental, and those with a quick intellect and a high level of motivation are likely to succeed, even though a simplistic analysis of physiological and biochemical findings might suggest that successful adjustment was improbable.

References

Abrahamson, M.J.; Wormald, P.J.; Millar, R.P.: Neuroendocrine regulation of thyrotropin release in cultured human pituitary cells. J. clin. Endocr. Metab. *65:* 1159–1163 (1987).

Ahlborg, J.; Felig, P.; Hagenfeldt, L.; Hendler, R.; Wahren, J: Substrate turnover during prolonged exercise in man. Splanchnic and leg metabolism of glucose, free fatty acids and amino acids. J. clin. Invest. *53:* 1080–1090 (1974).

Alekseev, Y.P.: Somatostatin: ego fiziologicheskoe znachenie i primenenie v klinike vnutrennikh boleznei (obzor literatury). Probl. Endokrinol. *23:* 93–96 (1977).

Aliev, M.A.; Lemeschko, V.A.; Bekbolotova, A.K.: Izmenenie gomeostatichaskogo balansa prostatsiklin-trombosangeneririuiushchikh sistem pri zoosotsial'nom stresse. Biull. Eksp. Biol. Med. *97:* 20–22 (1984).

Alimova, E.K., Astvatsaturyan, A.T.; Zharov, L.V.: Lipids and fatty acids in health and in some pathological states, p. 280 (Medicina, Moscow 1975).

Amir, S.; Brown, Z.W.; Amit, Z.: The role of endorphins in stress: evidence and speculations. Neurosci. Biobehav. Rev. *4:* 77–86 (1980).

Anastazi, A.: Psychological testing (in Russian), vol. 2, (Pedagogika, Moscow 1982).

Andronova, T.I.; Deryapa, R.N.; Solomatin, A.P.: Hemometeotropic reactions in healthy and sick subjects (in Russian), p. 248 (Medicina, Leningrad 1982).

Armstrong, W.E.; Sladek, C.D.; Sladek, J.R.: Characterization of noradrenergic control of vasopressin release by the organ-cultured rat hypothalamo-neurohypohyseal system. Endocrinology *111:* 273–279 (1982).

Arnoldi, I.A.: Acclimatization of man in the north and in the south (in Russian), p. 202 (Medicina, Moscow 1962).

Arshavskij, V.V.; Rotenberg, V.S.: Searching activity and adaptation (in Russian), p. 193 (Nauka, Moscow 1984).

Avtsyn, A.P.; Marachev, A.G.; Matveev, L.N.: The circumpolar hypoxia syndrome; in Proc 2nd All-Union Conf Human Adaptations to Various Geographic, Climatic and Industrial Conditions (in Russian), vol. 1, pp. 11–17 (Novosibirsk 1977).

Baevskyj, R.M.: Towards the problem of assessing the degree of stress of the regulatory systems of the organism. Adapatation and problems of general pathology (in Russian), vol. 1, pp. 44–48 (Novosibirsk 1974).

Baevskij, R.M.: Forecasting of states on the borderline of the norm and pathology (in Russian), p. 294 (Medicina, Moscow 1979).

Ball, J.H.; Kaminski, N.J.; Hardman, J.G.; Broadus, A.E.; Sutherland, E.W.; Liddle, G.W.: Effects of catecholamines and adrenergic blocking agents on plasma and urinary cyclic nucleotides in man. J. clin. Invest. *51:* 2124–2129 (1972).

Banister, E.W.; Griffiths, J.: Blood levels of adrenergic amines during exercise. J. appl. Physiol. *33:* 674–676 (1972).

Barbarash, N.A.; Dvurechenskaya, G.Y.: Physiology of adaptation processes (in Russian), pp. 251–302 (Nauka, Moscow 1986).

Barrett, A.I.: Lysosomal enzymes, lysozymes. A laboratory handbook; Ed. Ringle, J.T., Amsterdam, London. *219:* 46–49 (1972).
Basowitz, H.: Anxiety and stress. An interdesciplinary study of a life situation, p. 320 (McGraw-Hill, New York 1955).
Bass, D.E.: Electrolyte excretion during cold diureses (abstract). Fed. Proc. *13:* 8 (1954).
Baylic, P.H.; Zerbe, R.L.; Robertson, G.L.: Arginine vasopressin response to insulin- induced hypoglycemia in man. J. clin Endocr Metab. *53:* 935–940 (1981).
Bazhenov, A.F.; Posnyj, V.S.; Moshkin, M.P.: Circadian rhythms in man in the process of adaptation to the conditions of the far north; in Nepomnjashchih, Adaptations and problems of general pathology (in Russian), vol. 1, pp. 48–50 (Medical Academy of SSSR, Novosibirsk 1974).
Berezin, F.B.: Some aspects of psychic and psychological adaptation of man; in Symposium on psychological adaptation of man in the north, Vladivostok 1980 (in Russian), pp. 4–43 (Medical Academy of SSSR, Novosibirsk 1980).
Berezin, F.V.; Arshavskij, V.V.; Laneev, A.I., et al.: Results of electrophysiological studies on human adaptation and some psychophysiological correlations; in Symposium on psychological adaptation of man in the north, Vladivostok, 1980 (in Russian), pp. 97–133 (Medical Academy of SSSR, Novosibirsk 1980).
Bergelson, L.D.: Synthesis and research of prostaglandins (in Russian), p. 118 (Tallinn 1986).
Bobrov, N.I.; Lomov, O.P.; Tidhomirov, V.P.: Physiological aspects of acclimatization of man in the north (in Russian), p. 184 (Medicina, Leningrad 1979).
Bochkareva, E.V.: Correlation between prostaglandins and lipid metabolism disturbances in patients with coronary atherosclerosis (in Russian), Med. Sci. Diss. Moscow (1980).
Bolshakova, T.D.: Handbook of cardiology (in Russian), vol. 2, pp. 473–566 (Medicina, Moscow 1982).
Booth, M.A.; Thoden, J.S.; Reardon, F.D.; Jetté, M.; Rode, A,: The 1988 polar bridge expedition: counter-balancing the effects of changes in aerobic fitness and skiing economy on the relative stress of trekking. Can. J. Sport Sci. *14:* 103 (1989).
Borchgrevink, C.F.: Tests for capillary fragility and resistance disorders. Theory and methods; in Nils, Thrombosis and bleeding disorders, pp. 429–430. (Thieme, Stuttgart 1971).
Boriskin, V.V.: Life of man in the arctic and antarctic, in Matusov, Medical research on the arctic and antarctic expeditions (in Russian), pp. 50–65 (Hydrometeorological Publishing House, Leningrad 1973).
Bostic, J.; Kvetnansky, R.; Jansky, L.: Acta Univ. Carol. Biol. *3:* 257–261 (1979).
Bshir-Zader, T.S.; Malakhov, M.G.; Martens. V.I.; Chadov, V.I.: Psychological and physical adaptation in members of the USSR/Canada transarctic ski trek. 8th Int Congr Circumpolar Health, Whitehorse 1992, (in press).
Budd, G.M.: Acclimatization to cold in Antarctica as shown by rectal temperature response to a standard cold stress. Nature *193:* 886 (1962).
Bugard, P.; Albeaus-Fernet, M.; Romani, J.D. Rôle physiologique du système endocrien dans la fatigue. Med. Milit. fr *52:* 163–181 (1958).
Bunag, R.D.; Page, J.H.; McCubbin, J.: Neural stimulation of release of renin. Cardiovasc. Res. *1:* 67–75 (1967).
Burr, W.A.; Ramsden, D.B.; Griffiths, R.S.; Hofenberg, R,; Meinhold, H.; Wenzel, K.W.: Effect of a single dose of dexamethasone on serum concentrations of thyroid hormones. Lancet ii: 58–61 (1976).
Burton, A.C.: The average temperature of the tissues of the body. J. Nutr. *9:* 261–280 (1935).
Burton, A,C; Edholm, O.: Man in the cold. New York, Hafner, 1957.

Chowers, I.; Confortini, N.; Siegel, R.A: Inter-relationships between the central nervous system and patterns of adrenocorticotropic secretion following acute exposure to severe environmental conditions. Israel J. med. Scis *12:* 1010–1018 (1982).

Chubinskij, S.M.: Bioclimatology (in Russian), p. 199 (Medicina, Moscow 1965).

Cooper, K.E.: Mechanisms of human cold adaptation; in Shephard, Itoh, Circumpolar health, pp. 37–46 (University of Toronto Press, Toronto 1976).

Curtis, J.B.: Out in the cold: the physical and psychological hazards of living and working at isolated stations in the polar regions, p. 71 (Canadian Centre for Occupational Health and Safety, Ottawa 1985).

Danishevskij, G.M.: Human pathology and prevention of diseases in the north (in Russian), p. 412 (Medicina, Moscow 1968).

Davidenko, V.I.: Indices of aerobic performance efficiency and the contractile function of the myocardium in adaptation of man in the central antarctic; in Kaznacheev, Deryapa, Turchinsky, Clinical and experimental aspects of general pathology (in Russian), pp. 37–41 (Medical Academy of SSSR, Novosibirsk 1980).

Davis, J.R. Sheppard, M.C.: Mechanisms of TSH release: studies of forskolin, phorbol ester and A23187. Mol. cell. Endocrinol. *54:* 197–201 (1987).

Davis, T.R.: Chamber cold acclimatization in man. J. appl. Physiol. *16:* 1011–1015 (1967).

Davydova, N.; Tigranian, R.; Kalita, N.; Maldov, M.: The sympathetic-adrenomedullary, serotoninergic and histaminergic systems during a transarctic ski trek. Proc 8th Int Congr Circumpolar Health, Whitehorse 1992 (in press).

De Boer, A.C.; Mulder, H.; Fisher, H.R.; Schopman, W.; Hackeng, W.H.; Silberbusch, J.: Characteristic changes in the concentrations of some peptide hormones, in particular those regulating serum calcium, in acute pancreatitis and myocardial infarction. Acta med. scand. *209:* 193–198 (1981).

Deryapa, N.R.: Nature of the antarctic and human acclimatization (in Russian), p. 155 (Medicina, Moscow 1965).

Deryapa, N.R.; Ryabina, I.F.: Human adaptation in the polar regions of the earth (in Russian), p. 295 (Medicina, Leningrad 1977).

Dilman, V.M.; Nisevich, N.I.; Zubikova, I.I.; Romaniukha, A.A.; Moleva, T.P.: Endokrinnye sdvigi u detei s ostrym virusnym gepatitom. Probl. Endokrinol. *33:* 32–37 (1987).

Diprampero, P.E.: The energy cost of locomotion on land and in water. Int. J. Sports Med. *7:* 55–72 (1986)

Dluskaya, I.I.; Simonova, M.N., Veltischeva, I.E.: Lab. Delo *3:* 743–745 (1967).

Dorokova, B.; Chernikoskaya, T.; Roganov, A,; Kalita, N,; Melkonian, A,; Malakov, M.; Tigranian, R.; Shumulina, T.: The effects of prolonged exposure to arctic conditions on the somatotropin-somatostatin-somatomedin and parathyroid- calcitonin systems. Proc 8th Int Congr Circumpolar Health, Whitehorse 1992 (in press).

Drzhevetskaya, I.A.; Drzhevetsky, Y.M.: The hormonal regulation of the calcium turnover and secretory processes (in Russian), p. 132 (Nauka, Moscow 1983).

Duncan, D.E.: Multiple range and multiple F tests. Biometrics *11:* 1–42. (1955).

Duncombe, W.Y.: The colorimetric microdetermination of non-esterified fatty acids in plasma. Clin. chim. acta *9:* 122–125 (1964).

Dupuy, B.; Mounier, J.; Blanquet, P.: Quelques remarques sur les propriétés biologiques de la calcitonin. Une utilisation thérapeutique nouvelle possible. Anns Endocr *38:* 323–326 (1977).

Durnin, J.V.G.A.; Brockway, J.M.; Whitcher, H.W.: Effects of a short period of training of varying severity on some measures of physical fitness. J. Appl. Physiol. *15:* 161–165 (1960).

Durnin, J.V.G.A.; Womersley, J.A.: Body fat assessed from total body density and its estimation from skinfold thickness: measurements on 481 men and women aged 16 to 72 years. Br. J. Nutr. *32:* 77–97 (1974).
Eagan, C.J.: Local vascular adaptations to cold in man. Fed. Proc *22:* 947–951 (1963).
Eastman, C.J.; Lazarus, L.: Growth hormone release during sleep in growth retarded children. Archs Dis. Childh.*48:* 502–507 (1973).
Edholm, O.G.; Gunderson, K.E.: Polar human biology (Heinemann, London 1973).
Euler, U.S.; Lishajko, F.: The estimation of catecholamines in urine. Acta physiol. scand. *45:* 122–132 (1959).
Everitt, B.J.; Lightman, S.L.; Todd, K.: Brian stem noradrenergic pathways modulate vasopressin secretion in the rat. J. Physiol. Lond. *341:* 81 (1983).
Fabri, Z.I.; Pashchenko, A.E.: Tireoidnaia funktsiia u lits s gigerplaziei shchitovidnoi zhelezy v usloviiakh iodnoi nedostatochnosti. Probl. Endokrinol. *33:* 33–35 (1987).
Fitness and Amateur Sport: Canadian standardized test of fitness (CSTF): operations manual; 3rd ed., pp. 1–40 (Ministry of State, Fitness and Amateur Sport, Ottawa 1986).
Fletcher, B.L.; Dillard, C.T.; Tappel, A.L.: Measurement of fluorescent lipid peroxidation products in biological systems and tissues. Analyt. Biochem. *52:* 1–9 (1973).
Fortuine, R.: Circumpolar health '84 (University of Washington Press, Seattle 1985).
Forsius, H.: Pterygium, keratopathy, and pinguecula of the eyes in arctic and subarctic populations; in Shephard, Itoh, Circumpolar Health, pp. 364–373 (University of Toronto Press, Toronto 1976).
Fredrickson, R.C.; Burgis, V.; Harell, C.E.; Edwards, J.D.: Dual actions of substance P on nociception: possible role of endogenous opioids. Science *199:* 1359–1362 (1978).
Galbo, H.; Houston, M.E.; Christensen, N.J.; Holst, J.J.; Nielsen, B.; Nygaard, E.; Suzuki, J.: The effect of water temperature on the hormonal response to prolonged swimming. Acta physiol. scand. *105:* 326–337 (1979).
Geht, K.; Airapetyants, M.G.; Poppai, M.; et al.: Mechanisms of integrative brain activity, pp. 185–192 (Moscow 1981).
Gerasimova, T.I.: (in Russian), Lab. Delo *12:* 14–20 (1977).
Geselevich, V.A.: A medical handbook for the coach (in Russian), 2nd ed., p. 271 (Fizcultura i Sport, Moscow 1981).
Gibbs, D.M.: Dissociation of oxytocin, vasopressin and corticotropin secretion during different types of stress. Life Sci. *35:* 487–491 (1984).
Godin, G.; Shephard, R.J.: Activity patterns of the Canadian Eskimo; in Edholm, Gunderson, Polar human biology, (Heinemann, London 1973).
Goltseva, T.A.; Samodumova, M.G.; Dolgov, A.B.; Shvets, Z.A.: Laboratornaia diagnostika subklinicheskikh form giptireoza v èndemicheskom zobnom ochage. Probl. Endokrinol. *33:* 30–31 (1987).
Gordon, R.D.; Küchel, O.; Liddle, G.W.; Island, D.P.: Role of the sympathetic nervous system in regulating renin and aldosterone production in man. J. clin. Invest. *46:* 599–604 (1967).
Gotes, P.M.; Banham, D.R.: Bioassay of erythropoietin in mice made polycythemic by exposure. Nature, Lond. *191:* 1065 (1961).
Gray, G.D.; Smith, E.R.; Damassa, D.A.; Ehrenkranz, J.R.; Davidson, J.M.: Neuroendocrine mechanisms mediating the suppression of circulating testosterone levels associated with chronic stress in male rats. Neuroendocrinology *25:* 247–256 (1978),
Grigoriev, A.I.; Dorokhova, B.R.; Arzamazov, G.S.; Morukhov, B.V.: Ionoreguliuruiushchaia funktsiia pochek u cheloveka pri dlitelnykh komicheskikh poletakh i v modelnykh issledovaniiakh. Kosm. Biol. Aviakosm. Med. *16:* 29–33 (1982).
Gromova, E.A.: Serotonin and its role in the organism (in Russian), p. 197 (Nauka, Moscow 1966).

Gubachev, Y.M.; Iovlev, P.V.; Karvasarkij, B.D.; et al.: Emotional stress in the norm and pathology of man (in Russian), p. 224 (Medicina, Leningrad 1976).

Gubler, E.V.; Genkin, A.A.: The application of non-parametric statistical tests in medical and biological research (in Russian), p. 142 (Medicina, Leningrad 1973).

Guillemin, R.; Vargo, T.; Rossier, J.; Minick, S.; Ling, N.; Rivier, C.; Vale, W.; Bloom, F.: Beta endorphin and adrenocorticotropin are selected concomitantly by the pituitary gland. Science 197: 1365–1369 (1977).

Hammel, H.T.: Terrestrial animals in cold: recent studies of primitive man; in Dill, Handbook of physiology: adaptation to the environment, pp. 413–434 (American Physiological Society, Washington 1964).

Hampton, I.F.G.: Local acclimatization of the hands to prolonged cold exposure in the Antarctic. Br. Antarctic Survey Bull. 19: 9–56 (1969).

Hardy, J.D; Dubois, E.F.: The technic of measuring radiation and convection. J. Nutr. 15: 461–475 (1938).

Hart-Hansen, J.P.; Harvald, B.: Circumpolar health '81 (Nordic Council for Arctic Medical Research, Copenhagen 1981).

Hatch, F.T., Lees, R.S.: Practical methods for plasma lipoprotein analysis. Adv. Lipid Res. 6: 2–68 (1968).

Hensel, H.; Schäfer, K.: Static and dynamic activity of cold receptors after long exposure to various temperatures. Pflügers Arch 392: 291–294 (1982).

Hesse, B.; Nielsen, I.: Suppression of plasma renin activity by intravenous infusion of antidiuretic hormone in man. Clin. Sci. mol. Med. 52: 357–364 (1977).

Hildes, J.A.; Schaefer, O.; Sayed, J.E.; Fitzgerald, E.J.; Koch, E.A.: Chronic lung disease and cardiovascular consequences in Iglooligmuit; in Shephard, Itoh, Circumpolar health, pp. 327–337 (University of Toronto Press, Toronto 1976).

Hinckel, P.; Schröder-Rosenstock, K.: Responses of pontine units to skin temperature changes in the guinea pig. Pflügers Arch 392: 344–346 (1982).

Hiramatsu, K.; Yamada, T.; Katakura, M.: Acute effects of cold on blood pressure renin-angiotensin-aldosterone system, catecholamines and adrenal steroids in man. Clin. exp. Pharmacol Physiol. 2: 171–179 (1984).

Horky, K.; Rojo-Ortega, J.M.; Rodriguez, J.; Boucher, R.; Genest, J.: Renin, renin substrate and angiotensin I converting enzyme in the lymph of rats. Am. J. Physiol. 220: 307–311 (1971).

Ikeda, T.; Takeuchi, T.; Ito, Y.; Murakami, I.; Mokuda, O.; Tominaga, M.: Mashiba, H.: Effect of thyrotropin on conversion of T_4 to T_3 in perfused rat liver. Life Sci. 38: 1801–1806 (1986).

Ilyin, E.A.: Changes in the nervous-psychic state and medicamentous methods of its normalization in the conditions of wintering at the 'Vostok' Station; in Matusov, AANII Proc medical research in the arctic and antarctic expeditions (in Russian), vol. 2, pp. 90–98 (Gidrometeoizdat, Leningrad 1971).

Ivanov, R.P.: Bioenergetics and temperature homeostasis (in Russian), p. 169 (Medicina, Leningrad 1972).

Jessen, R.: An assessment of human regulatory non-shivering thermogenesis. Acta anaesth scand. 24: 138–143 (1980).

Jetté, M.: Guide for anthropometric measurements of Canadian adults, pp. 1–16 (Claude Tessier Gestion, Montreal 1983a).

Jetté, M.: Anthropometric characteristics of the Canadian population. Nutrition Canada Survey, 1970–72, pp. 1–69 (Nutrition Canada, Ottawa 1983b).

Jetté, M.; Rode, A.; Koscheyev, V.; Booth, M.; Thoden, J.: Anthropometric and fitness characteristics of members of the USSR/Canada Polar Bridge Expedition. Can. J. Sport Sci. 14: 119 (1989).

Jetté, M.; Sidney, K.H.; Kimick, A.: Comparison of hand grip and push dynamometers as indicators of body strength. Can. J. Sport Sci. *9:* 22 (1984).

Jimenez, J.; Zuniga-Guajardo, S.; Zinman, B.; Angel, A.: Effects of weight loss in massive obesity on insulin and C-peptide dynamics: sequential changes in insulin production, clearance and sensitivity. J. clin. Endocr Metab. *64:* 661–668 (1987).

Jones, N.; Kane, M.: Inter-laboratory standardization of methodology. Med. Sci. Sports *11:* 368–372 (1979).

Kalita, N.; Krylov, Y.F.; Tigranian, R.A.; Daydova, N.A.: Stress. The role of catecholamines and other neurotransmitters; in Usdin Kvetnansky, Axelrod, Stress, vol. 2. pp. 967–973 (New York 1984).

Kalita, N.; Krylov, Y.; Tigranian, R.A.; Davydova, N.: 4th Symposium on Catecholamines and Other Neurotransmitters in Stress (in Russian) Smolenice Castle 1987, (abstract). p. 42

Kalita, N.; Tigranian, R.; Dorokova, B.; Shumulina, T.; Malakov, M.; Chernikovskaya, T.; Roganov, A.: The hypophysis-thyroid gland system during exposure to extreme arctic conditions. Proc 8th Int Congr. Circumpolar Health, Whitehorse 1992a (in press).

Kalita, N,; Tigranian, R.; Shumulina, T., Dorokova, B.; Roganov, A.; Melkonian, A.; Malkov, M.; Chernikovskaya, T.: Pituitary adrenocortical activity during a transarctic ski trek. Proc 8th Int Congr Circumpolar Health, Whitehorse 1992b, (in press).

Kaluzenko, R.K.: The coagulating system of the blood in polar explorers of the 'Vostok' Station; in Matusov, AANII Proc medical research in the arctic and antarctic expressions (in Russian), vol. 2, pp. 172–176 (Gidrometoizdat, Leningrad 1971).

Kamberi, I.A.; Mical, R.S.; Porter, J.C.: Effect of anterior pituitary perfusion and intraventricular injection of catecholamines on prolactin release. Endocrinology *89:* 1042–1046 (1971).

Kandror, I.S.: Essays on physiology and hygiene of man in the far north (in Russian), p. 280 (Medicina, Moscow 1968).

Karlson, L.; Levi, L.; Ure, L.: in Proc Int Symp Emotional Stress (in Russian), pp. 152–160 (Leningrad 1970).

Katsia, G.V.; Chirkov, A.M.; Goncharov, N.P.: Vliianie Liuliberina i khorionicheskogo gonadotropina na uroven luiteiniziruiushchego gormona i testosterona v krovi u obezian v usloviiakh ostrogo stressa. Probl. Endokrinol. *30:* 73–77 (1984).

Kaznacheev, V.P.: Topical issues of human adaptation; in Vasilyev, Climatomedical problems and issues of medical geography of Siberia. A collection of papers, vol. 1, pp. 6–18 (Medical Academy of SSSR, Novosibirsk 1974).

Kaznacheev, V.P.: The relationship between physiological and biological mechanisms of human adaptation; in, Shephard, Itoh, Circumpolar health, pp. 23–37 (University of Toronto Press, Toronto 1976).

Kaznacheev, V.P.; Egunova, M.M; Kulikov, V.Y.; Kim, L.B.; Molchanova, L.V.; Stuhljaev, V.A.; Kolosova, N.G.; Kolesnikova. L.I.: Metabolism and peroxidation of lipids in man at adaptation to the conditions of the far north; in Derjapa, Topical issues of human adaptation in the far north and Antarctica (in Russian), pp. 3–15 (Medical Academy of SSSR, Novosibirsk 1976).

Kaznacheev, V.P.; Kulikov, V.J.; Soini, E.; Leppaluoto, J.; Stenback, F.: Bibliography on arctic medical research in the USSR. Arctic med Res *39:* 11–140 (1985).

Keatinge, W.R.: The Effect of general chilling on the vasodilator response to cold. J. Physiol., Lond. *139:* 497–507 (1957).

Khaskin, V.V.: Energetics of heat formation and cold adaptation (in Russian) (Nauka, Novosibirsk 1975).

Khokhar, A.M.; Slater, I.D.; Forsling, M.L.; Payne, N.N.: Effect of vasopressin on plasma volume and renin release in man. Clin. Sci. mol. Med. *50:* 415–424 (1976).

Knepel, W.; Nutto, D.; Hertting, G.: Evidence for inhibition by β-endorphin of vasopressin release during foot-shock induced stress in the rat. Neuroendocrinology 34: 353–356 (1982).

Knepel, W.; Przewlecki, R.; Nutto, D.; Herz, A.: Foot-shock stress-induced release of vasopressin in adrenohypophysectomized and hypophysectomized rats. Endocrinology 117: 292–299 (1985).

Korjranovskij, B.B.: Exposure and over-exposure to cold and their prevention (in Russian), p. 247 (Medicina, Leningrad 1966).

Korolenko, T.P.: Psychophysiology of man in the extreme conditions (in Russian), p. 272 (Medicina, Leningrad 1978).

Koscheyev, V.S.; Lartzev, M.A.; Martens, V.K.: Psychological adaptation in participants of a transarctic ski trek. Proc 8th Int Congr Circumpolar Health, Whitehorse 1992, (in press).

Kosovskii, M.I.; Mirakhmedov, M.M.; Katkova, S.P.; Makhkamova, R.U.: Osobennosti narusheniia uglevodnago obmena pri stresse. Probl. Endokrinol. 34: 48–51 (1988).

Kotchen, T.A.; Roy, M.W.: in Dunn, Renal endocrinology, pp. 241–271 (Medicina, Moscow 1987).

Kozhemyakin, L.A.; Korostovtsev, D.S.; Koroleva, T.R.: Tsillicheskii adenozin-3'5'-monofosfat v organakh i tkaniakh v prostresse adaptatsii organizma k ekstremalnym vozdeistviiam. Biull. Eksp. Biol. Med. 84: 567–568 (1977).

Krylov, Y.; Tigranian, R.A.: Stress, adaptation and functional disturbances. Proc 1st Symp Stress, Adaptation and Functional Disturbances (in Russian), Kishinev 1984, p. 116.

Krylov, Y.; Tigranian, R.A: Gormonal no-metabolicheskii status organizma cheloveka v usolviiakh Krainego Severa. Kosmich. Biol. Aviakosm. Med. 20: 85–88 (1986).

Krylov, Y.; Tigranian, R.A.; Pavlova, E.; Chudnovskaya, E.: Cyclic nucleotides and the system that regulates enzymatic reactions (in Russian), pp. 69–70 (Ryazan 1985).

Kryshanovskii, G.N.; Igonkina, S.I.; Trubetskaia, V.V.; Oeme, P.; Binert, M.: Vlianie substantsii P i ee fragmentov na fiziologicheskuiu i patologicheskuiu bol. Biull. Eksp. Biol. Med. 105: 655–657 (1988).

Kulikov, V.Y.; Lyakhovich, V.V.; Reactions of free radical oxidation of lipids and some indices of oxygen metabolism, in Mechanisms of human adaptation in high latitudes (in Russian), vol. 4, pp. 60–87 (Medicina, Leningrad 1980).

Kulinsky, V.I.; Plotnikov, N.Y.: (abstract). Congr. Physiologically active compounds for medicine, Yerevan 1982, pp. 345–346.

Kvetnansky, R.; Gewirtz, I.P.; Weise, V.K.; Kopin, I.J.: Catecholamine synthesizing enzymes in the rat adrenal gland during exposure to cold. Am. J. Physiol. 220: 928–931 (1971).

LeBlanc, J.A.: Evidence and meaning of acclimatization to cold in man. J. appl. Physiol. 9: 595–598 (1956).

LeBlanc, J.A.; Hildes, J.A.; Heroux, O.: Tolerance of Gaspé fishermen to cold water. J. appl. Physiol. 15: 654–658 (1960).

Leppäluoto, J.; Korhonen, I.; Huttunen, P.; Hassi, J.: Serum levels of thyroid and adrenal hormones, testosterone, TSH, LH, GH and prolactin in man after a 2 h stay in a cold room. Acta physiol. scand. 132: 543–548 (1988).

Linderholm, H.: Circumpolar Health '87 (Nordic Council for Arctic Medical Research, Copenhagen 1989).

Livingstone, S.D.: Changes in cold-induced vasodilatation during Arctic exercises. J. appl. Physiol. 40: 455–457 (1976a).

Livingstone, S.D.: Effect of vitamin C on cold-induced vasodilatation. Lancet ii: 319 (1976b).

Livingstone, S.D.; Nolan, R.W.; Keefe, A.A.: The effects of a 90 day polar ski expedition of cold acclimatization. Proc 8th Int Congr Circumpolar Health, Whitehorse 1992, (in press).

Losev, N.I.; Khitrov, N.K.; and Grachev, S.V.: Pathophysiology of extreme states. Handbook (in Russian), p. 72 (Moscow 1988).

Lowry, O.H.; Rosebrough, N.J.; Farr, A.L.; Randall, R.J.: Protein measurement with the Folin phenol reagent. J. biol. Chem. *193:* 265–275 (1951).

Loskutova, T.D.: Assessment of the functional state of the CNS of man by the parameters of a simple motor reaction (in Russian). USSR Physiol. J. *1:* 3–12 (1975).

Malakhov, M.I.; Tigranian, R.A.; Khelevsky, Y.I.; et al.: Adaptation of organisms to natural conditions (abstract, in Russian). Syktyvkar *3:* 34 (1982)

Marshak, M.E.: Physiological principles of hardening of human organisms (in Russian), p. 150 (Medicina, Leningrad 1965).

Martynenko, F.P.; Rozhko, O.T.; Novikova, N.P.: Vklluchenie leitsina-C14 v somatotropnyi gormon krys v zavisimosti ot urovnia tireoidnykh gormonov v organizme. Probl. Endokrinol. *20:* 107–110 (1974).

Mashford, M.L.; Nilsson, G.; Rokäeus, A., Rosell, S.: The effect of food ingestion on circulating neurotensin-like immunoreactivity (NLTI) from the gut in anaesthetized dogs. Acta physiol. scand. *104:* 375–379 (1978).

Matlina, E.S.; Bolshakova, T.D.; Shirinian, E.A.: Methods of clinical biochemistry of hormones and mediators (in Russian), vol. 2, pp. 60–67 (Medgiz, Moscow 1974).

Matusov, A.L.: The conditions of life and the general state of the participants of polar expeditions (in Russian), p. 232 (Gidrometeoizdat, Leningrad 1979).

Mayer, G.P.; Hurst, J.C.; Barto, J.A.; Keaton, J.A.; Moore, M.P.: Effect of epinephrine on parathyroid hormone secretion in calves. Endocrinology *104:* 1181–1187 (1979).

McNeilly, A.S.: Prolactin and the control of gonadotrophin secretion. J. Endocr *115:* 1–5 (1987).

Meehan, J.P.: Racial and individual differences in the peripheral vascular response to a cold stimulus. Am. J. Physiol. *179:* 547–561 (1954).

Meerson, F.Z.: Theoretical problems of low temperature influence the organism: total hypothermia (in Russian), part 2, pp. 124–132 (Vladimir, 1972).

Meerson, F.Z.: Pathogenesis and prevention of heart damage induced by stress and ischemia (in Russian), p. 269 (Medicina, Moscow 1984).

Melkonian, A.; Kalita, N.; Tigranian, R.; Malakhov, M.; Dorokova, B.; Chernikovskaya, T.; Roganov, A.; Shumulina, T.: The Effects of extreme conditions on the pituitary-gonadal system during a transartic ski trek. 8th Int Congr Circumpolar Health, Whitehorse 1992, (in press).

Mest, H.J.; Sziegoleit, V.; Forster, W.: Vliianie stressovykh nagruzok na kontsentratsiiu prostaglandinov v plasme krovi studentov v period ékzaminov. Farmakol. Toksidol. *45:* 110–112 (1982).

Mevkh, A.T.: Production and application of biocatalysts in national economy and in medicine (in Russian), part 1, pp. 37–38 (Moscow 1983).

Michajlovskij, N.; Lichardus, B.; Kvetnansky, R.; Ponec, J.: Effect of acute and repeated immobilization on food and water intake, urine output and vasopressin changes in rats. Endocrinol. exp. *22:* 143–157 (1988).

Mills, D.E.; Robertshaw, D.J.: Response of plasma prolactin to changes in ambient temperature J. clin. Endocr. Metab. *2:* 279–283 (1981).

Mills, D.E.; Robertshaw, D.1: Plasma prolactin responses to acute changes in central blood volume in man. Hormone Res. *18:* 153–159 (1983).

Mills, F.J.: The endocrinology of stress. Aviat. Space environ. Med. *56:* 642–650 (1985).

Mirakhmedov, M.: The role of iodothyronine metabolism in the regulation of the thyroid status of the organism: Clinical and experimental research (in Russian); Med. Sci. diss. Tashkent University, Tashkent (1979)

Miroshnikov, M.P.: One conception of psychic stress according to data of foreign studies, in Alatortsev, Problems of sport psychology: a collection of works (in Russian), vol. 1, pp. 137–165 (Moscow 1971).

Mishina, N.F.: Calcitonin participation in stress development in postnatal ontogenesis (in Russian); Med. Sci. diss. University of Leningrad, Leningrad (1981).

Morganroth, J.; Maron, B.J.: The athlete's heart syndrome. A new perspective. Ann. N.Y. Acad. Sci. *301:* 931–939 (1977).

Morton, A.D.; Fitch, K.D.: Exercise-induced bronchial obstruction; in Torg, Welsh, Shephard, Current therapy in sports medicine, vol. 2 (Decker, Burlington 1990).

Morton, J.J.; Garcia del Rio, C.; Hughes, M.J.: Effect of acute vasopressin infusion on blood pressure and plasma angiotensin II in normotensive and DOCA-salt hypertensive rats. Clin. Sci. *62:* 143–149 (1982).

Mostovoy, V.S.: Adaptation of organisms to natural conditions (abstract, in Russian). Syktyvkar *3:* 40 (1982). (Manus 10, p 18).

Murray, S.J.; Shephard, R.J.; Greaves, S.; Allen, C.; Radomski, M.: Effects of cold stress and exercise on fat loss in females. Eur. J. appl. Physiol *55:* 610–618 (1986).

Naumenko, E.V.; Popova, I.K.: Serotonin and melatonin in regulation of the endocrine system (in Russian), p. 164 (Nauka, Novosibirsk 1975).

Nebolsina, L.I.; Poleshchuk, V.S.; Markhov, K.M.: O vzaimodeistvii prostaglandinov i simpatiko-adrenalovoi sistemy. Patol. Fiziol. Eksp. Ter. *6:* 48–50 (1982).

Nelms, J.D.; Soper, D.J.G.: Cold vasodilatation and cold acclimatization in the hands of British fish filleters. J. appl. Physiol. *17:* 346–348 (1962).

Nikolsky, N.N.; Troshin, A.A.: Transport of sugars through cellular membranes (in Russian), p. 222 (Nauka, Leningrad 1973).

Nutrition Canada: Food consumption patterns report (Health and Welfare Canada, Ottawa 1976).

Nutrition Canada: Recommended nutrient intake for Canadians (Nutrition Canada, Ottawa 1983).

Oehme, P.; Hecht, K.; Piesche, L,; Hilse, H.; Morgenstern, E.: Substance P as a modulator of physiological and pathological processes; in Mersan, Traczyk, Neuropeptides and neural transmission, pp. 73–84 (Raven Press, New York 1980a).

Oehme, P.; Hilse, H.; Morgenstern, E.; Göres, E.: Substance P: does it cause analgesia or hyperalgesia? Science *208:* 305–307 (1980b).

Oehme, P.; Hilse, H.; Görne, R.C.; Hecht, K.: Influence of substance P on nociception and stress. Pharmazie *40:* 568–570 (1985).

O'Hara, W.; Allen, C.; Shephard, R.J.: Loss of body fat during an arctic winter expedition. Can. J. Physiol. *55:* 1235–1241 (1978).

O'Malley, B.P.; Cook, N.; Richardson, A.; Barnett, D.B.; Rosenthal, F.D.: Circulating catecholamines, thyrotropin, thyroid hormones and prolactin responses of normal subjects to acute cold exposure. Clin. Endocrinol. *21:* 285–291 (1984).

Onaka, T.; Hamamura, M.; Yagi, K.: Potentiation of vasopressin secretion by foot-shocks in rats. Jap. J. Physiol. *36:* 1253–1260 (1986a).

Onaka, T,; Hamamura, M.; Yagi, K.: Suppression of vasopressin secretion by classically conditioned stimuli in rats. Jap. J. Physiol. *36:* 1261–1266 (1986b).

Panin, L.E.: Modern problems of the respiratory biochemistry and clinic (in Russian), pp. 52–56 (Ivanovo 1970).

Panin, L.E.: Some medical and biological aspects of processes of adaptation (in Russian), pp. 34–43 (Novosibirsk 1975).

Panin, L.E.: Energetic aspects of adaptation (in Russian), p. 190 (Medicina, Leningrad 1978).

Panin, L.E.: Biochemical mechanisms of stress (in Russian), p. 234 (Nauka, Novosibirsk 1983).
Panin, L.E.: Abstracts 15th Congr Pavlov's All-Union Physiological Society (in Russian), vol. 1, p. 26 (Kishinev, 1987).
Panin, L.E.; Moshkin, M.P.; Shevchencko, Y.S.: Energy aspects of human adaptation in high latitudes; in Kaznacheev, Derjapa, Turchinsky, Problems of ecology of man in the far north (in Russian), pp. 9–17 (Medical Academy of SSSR, Novosibirsk 1979).
Panin, L.E.; Mayanaskaya, N.N.; Borodin, A.A.: Changes in blood acid hydrolase activity during a transarctic ski trek. Proc 8th Int Congr Circumpolar Health, Whitehorse 1992b (in press).
Panin, L.E.; Kunitsin, V.G.; Nekrasova, M.F.; Kolosova, N.G.: Structural and functional features of erythrocyte membranes in members of a transarctic ski trek. Proc 8th Int Congr Circumpolar Health, Whitehorse 1992a, (in press).
Panin, L.E.; Ostanina, L.S.; Tretyakova, T.A.; Kolpakov, A.R.: Energy metabolism changes in participants of a transarctic ski trek. Proc 8th Int Circumpolar Health, Whitehorse 1992c, (in press).
Panin, L.E.; Vloshinsky, P.E.; Kolosova, I.E.; Kolpakov, A.R.: Changes in endocrine activity in participants of a transarctic ski trek. Proc 8th Int Congr Circumpolar Health, Whitehorse 1992d, (in press).
Pastukhov, Y.P.; Khaskin, V.V.: (in Russian). Uspekhi Physiol. Nauk. *10:* 121–142 (1979).
Pecile, A.; Müller, E.: Suppressive action of corticosteroids on the secretion of growth hormone. J. Endocr. *36:* 401–408 (1966).
Poggenpol V.S.; Ilyin, E.A.: Changes in some aspects of the oxygen regime of human organisms at protracted stay in the extreme conditions; in Matusov, AANII Proc. Medical research in the arctic and antarctic expeditions, vol. 2, pp. 182–186 (Gidrometeoizdat, Leningrad 1971).
Produtskij, P.A.; Vorobyev, A.A.: Peculiarities of the acclimatization processes in seamen of the transpolar regions; Acclimatization of man in the polar regions, pp. 64–65 (n: Leningrad 1969).
Queener, S.F.; Bell, N.H.; Larson, S.M.; Henry, D.P.; Slatopsky, E.: Comparison of the regulation of calcitonin in serum of old and young buffalo rats. J. Endocr. *87:* 73–80 (1980).
Radomski, M.W.; Boutelier, C.: Hormone response of normal and cold-preadapted humans to continuous cold. J. appl. Physiol. *53:* 610–616 (1982).
Reid, I.A.; Morris, B.J.; Ganong, W.F.: The renin-angiotensin system. A. Rev. Physiol. *40:* 377–409 (1978).
Repcedova, D.; Mikulaj, L.: Plasma testosterone response to HCG in normal men without and after administration of anabolic drug. Endokrinologie *69:* 115–118 (1977).
Rivolier, J.; Goldsmith, R.; Lugg, D.J.; Taglor, A.J.W.: What adaptations developed?; in Goldsmith, R. (Ed.): Man in the Antarctic, pp. 105–148 (Taylor and Francis, Philadelphia 1988).
Rode, A.; Shephard, R.J.: Lung function in a cold environment. A current perspective; in Fortuine, Circumpolar health '84 pp. 60–63 (University of Washington Press, Seattle 1985).
Roganov, A.; Kalita, N.; Tigranian, R.; Dorokova, B.; Chernikovskaya, T.; Malkov, M.; Shumulina, T.; Melkonian, A.: The effect of severe cold and exercise on the activity of the renin-angiotensin-aldosterone system. Proc. 8th Int. Congr Circumpolar Health, Whitehorse 1992, (in press).
Ross, R.J.; Borges, F.; Grossman, A.; Smith, R.; Ngahfoong, L.; Rees, L.H.; Savage, M.O.; Besser, G.M.: Growth hormone pretreatment in man blocks the response to growth-hormone releasing hormone: evidence for a direct effect of growth hormone. Clin. Endocrinol. *26:* 117–123 (1987).

Roy, C.; Hall, D.; Karish, M.; Ausiello, D.A.: Relationship of (8-lysine) vasopressin receptor transition to receptor functional properties in a pig kidney cell line (LLC-PK$_1$). J. biol. Chem. *256:* 3423–3430 (1981).

Saakov, B.A.; Eremina, S.A.; Gulyants, E.S.: Biull. Eksp. Biol. Med. *68:* 25–28 (1969).

Said, S.I.: Prostaglandins and the lung. Bull. Eur. Physiopath. Respir. *17:* 487–488 (1981)

Scapagnini, U.; Preziosi, P.: Role of brain nordarenaline in the tonic regulation of hypothalamic hypophyseal adrenal axis. Prog. Brain Res. *39:* 171–184 (1973).

Scholander, P.F.; Hammel, H.T.; Andersen, K.L.; Loyning, Y.: Metabolic acclimation to cold in man. J. appl. Physiol. *12:* 1–8 (1958).

Schonbaum, E.; Sellers, E.A.; Johnson, G.E.: Noradrenaline and survival of rats in a cold environment. Can. J. Biochem. *41:* 975–983 (1963).

Scriven, A.J.; Brown, M.J.; Murphy, M.B.; Dollery, C.T.: Changes in blood pressure and plasma catecholamines caused by tyramine and cold exposure. J. cardiovasc. Pharmacol. *6:* 954–960 (1984).

Sellers, E.A.: Adaptive and related phenomena in rats exposed to cold. A review. Rev. Can. Biol. *16:* 175–188 (1957).

Seliatitskaia, V.G.; Solenov, E.I.; Shorin, Y.P.; Ivanova, L.N.: Pokazateli gormonalnoi reguliatsii vodno-solevogo obmena i retseptsiia tsAMF v sosochke pochki krys pri adaptatsii k kholodu. Biull. Eksp. Biol. Med. *100:* 393–394 (1985).

Semenov, V.I.: Materials on disease incidence in the young during the first months of their stay in the North and problems of disease prevention; in Derjapa, Topical issues of human adaptation in the far north and the Antarctic (in Russian) pp. 36–45 (Medical Academy of USSR, Novosibirsk 1976).

Shchedrin, A.S.: The peculiarities of the flu epidemics and the seasonal dynamics of acute respiratory diseases in the north; in Derjapa, Topical issues of human adaptation in the far north and Antarctica (in Russian), pp. 45–49. (Medical Academy of USSR, Novosibirsk 1976).

Shcherbakova, V.S.; Rom-Boguslavskaya, E.S.: Osobannosti reaktsii shchitovidnoi zhelezy krys na melatonin u pinealéktomirovannykh krys. Probl. Endokrinol. *34:* 75–78 (1988).

Shephard, R.J.: Physiology and biochemistry of exercise (Praeger, New York 1982).

Shephard, R.J.: Biochemistry of Exercise (C.C. Thomas, Springfield 1983).

Shephard, R.J.: Adaptation to exercise in the cold. Sports Med *2:* 59–71 (1985).

Shephard, R.J.: Exercise and the Immune system. Can. J. Sport Sci. (in press, 1991).

Shephard, R.J.; Itoh, S.: Circumpolar health (University of Toronto Press, Toronto 1976).

Shevchencko, Y.S.; Moshkin, A.F.; Saenko, M.P.: Adaptational changes of the cardio-respiratory system in polar explorers during 1-year long activity of the antarctic expedition; in Derjapa, Topical issues of human adaptation in the far north and the antarctic (in Russian), pp. 49–57 (Medical Academy of USSR, Novosibirsk 1976).

Shumulina, T.; Tigranian, R.; Kalita, N.; Chernikovskaya, T.; Dorokova, B.; Roganov, A.; Melkonian, A.; Malkov, M.: The effects of severe cold and exercise on blood levels of cyclic nucleotides and prostaglandins. Proc. 8th Int. Congr. Circumpolar Health, Whitehorse 1992; (in press).

Simmonds, M.A.: Inhibition by atropine of the increased turnover of noradrenaline in the hypothalamus of rats exposed to cold. Br. J. Pharmacol. *41:* 224–229 (1971).

Siri, W.E.: The gross composition of the body. Adv. biol. med. Phys. *4:* 239–280 (1956).

Skorobogatova, A.M.; Paramonov, Y.A.; Lukachev, V.V.: Changes in the frequency and character of respiratory movements in humans at low temperatures; in Acclimatization of man in the polar regions (in Russian), pp. 100–102 (Leningrad 1969).

Skulachev, V.P.: The energy transformation in biomembranes (in Russian), p. 204 (Nauka, Moscow 1982).

Slepushkin, V.D.; Stepnoi, P.S.: Izmaniia pohazatelei vodno-solovogo ravnovesiia pri sochetanii obshchego i mestnogo okhlazhdeniia. Ortop. Travmatol. Protez. Nov *ii:* 30–33 (1980).
Slonim, A.D.: Physiological adaptation to heat and cold (in Russian), p. 126 (Nauka, Leningrad 1969).
Smelik, P.G.: in Usdin, Kvetnansky, Axelrod, Catecholamines and stress, Oxford vol. 1, pp. 18–22 (Pergamon Press, New York 1984).
Smith, D.F.; de Long, W.: Pharmacopsychiatrie *8:* 132–135 (1975). (Manus 10, p 18).
Smith, M.J.; Cowley, A.W.; Guyton, A.C.; Manning, R.D.: Acute and chronic effects of vasopressin on blood pressure, electrolytes and fluid volumes. Am. J. Physiol. *237:* 232–237 (1979a).
Smith, P.H.; Porte, D.; Robertson, R.P.: b Endocrine pancreas and diabetes; in Pierluissi, pp. 64–65 (Excerpta Medica, Amsterdam 1979a).
Sowers, J.R.; Carlson, H.E.; Brautbar, N.; Hershman, J.M.: Effect of dexamethasone on prolactin and TSH responses to TRH and metclopramide in man. J. clin. Endocr Metab. *44:* 237–241 (1977a).
Sowers, J.R.; Raj, R.P.; Hershman, J.M.; Carlson, H.E.; McCallum, R.W.: The effect of stressful diagnostic studies and surgery on anterior pituitary hormone release in man. Acta endocr., Copenh. *86:* 25–32 (1977b).
Sproule, J.; Jetté, M.; Rode, A.: Medical observations on members of the USSR/Canada Polar expedition. Can. J. Sport Sci. *14:* 137 (1989).
Stabrovskii, E.M.; Korovin, K.F.: Vliianie vozdushogo okhlazhdeniia na funktsiiu simpatoadrenalovi systemy u krys. Fyziol. J. SSSR *57:* 539–545 (1971).
Stabrovsky, E.M.; Korovin, K.F.: Katekholaminy v tkaniakh krys i ikh obmen pri okhlazhdenii. Fyziol. J. SSSR *58:* 414–418 (1972).
Starkova, N.T.: in Dagett, Clinical endocrinology, pp. 82–105 (Medicina, Moscow 1981).
Tanaka, M.: Experimental studies on human reaction to cold. Bull. Tokyo Med. Dent Univ. *18:* 169–177 (1971).
Tarkhan, A.U.: Assessment of the nervous psychic state in polar explorers in climate of the arctic; in Derjapa, Topical issues of human adaptation in the far north and the Antarctic (in Russian), pp. 96–103 (Medical Academy of USSR, Novosibirsk 1976).
Taylor, A.L.; Davis, B.B.; Pawlson, L.I.; Josimovich, J.B.; Mintz, D.H.: Factors influencing the urinary excretion of $3'5'$ adenosine monophosphate in humans. J. clin. Endocr. Metab. *30:* 316–324 (1970).
Taylor, S.L.; Lamden, M.P.; Tappel, A.L.: Sensitive fluorimetric method for tocopherol analysis. Lipids *11:* 530–538 (1976).
Tigranian, R.A.; Orloff, L.L.; Kalita, N.F.; Davydova, N.A.; Pavlova, E.A.: Changes of blood levels of several hormones, catecholamines, prostaglandins, electrolytes and cAMP in man during emotional stress. Endocrinol. exp. *14:* 101–112 (1980).
Tigranian, R.A.; Krylov, Y.; Kalita, N.: Peptide and monoamine neurohormones in neuroendocrine regulation (abstract), Congr. Biochemistry, Leningrad, 1985, p. 137.
Tigranian, R.A.; Krylov, Y.; Kalita, N.: (abstract). 3rd Int. Conf. combined effects of environmental factors, Tampere 1988b, p. 68.
Tigranian, R.A.; Kalita, N.; Malakhov, M.: (abstract). Symp. human biometeorology, Bratislava 1988a, p. 81.
Tigranian, R.; Kalita, N.; Malakov, M.; Shumulina, T.; Chernikovskaya, T.; Roganov, A.; Melkonian, A.; Dorokova, B.: The blood content of neuropeptides in skiers during a transarctic ski trek. Proc. 8th Int. Congr. Circumpolar Health, Whitehorse 1992, (in press).
Tikhorimov, I.I.: Bioclimatology of the central antarctic and acclimatization of man (in Russian), P. 197 (Nauka, Moscow 1968).

Tretyakova, T.A.; Panin, L.E.: Determination of key steps of glycolysis in tissues with different functional specialization; in Nepomnyaschikh, New methods of scientific research, diagnosis and treatment, pp. 7–9 (USSR Academy of Sciences, Novosibirsk 1978).

Tsibizov, G.G.; Hormonal regulation of the calcium and phosphate homeostasis during physical activity. Physiol. J. USSR 65: 1539–1543 (1979).

Udintsev, N.N.; Fedorova, T.S.; Sersbrennikova, I.A. et al.: in Biochemical ecology. Experimental and clinical biochemistry (in Russian), pp. 145–148 (Sverdlovsk 1985).

Urbach, V.I.: Statistical analysis in biological and medical research (in Russian), p. 296 (Medicina, Moscow 1975).

Utehin, B.A.; Malakhov, M.G.: Thermoregulation and heat exchange in participants of a transarctic ski trek. 8th Int. Congr. Circumpolar Health, Whitehorse 1992, (in press).

Vander, A.J.: Inhibition of renin release in the dog by vasopressin and vasotocin. Circulation Res. 23: 605–609 (1968).

Vandongen, R.: Inhibition of renin secretion in the isolated rat kidney by antidiuretic hormone. Clin. Sci. mol. Med. 49: 73–76 (1975).

Van Someren, R.N.; Coleshaw, S.R.; Mincer, P.J.; Keatinge, W.R.: Restoration of thermoregulatory response to body cooling by cooling hands and feet. J. appl. Physiol. 53: 1228–1233 (1982).

Varfolomeev, S.D.; Mevkh, A.T.: Prostaglandins as molecular bioregulators: biokinetics, biochemistry, medicine (in Russian), p. 307, (Nauka, Moscow 1985).

Vasilevskij, N.N.; Soroko, S.I.; Bogoslovskij, M.M.: Psychological aspects of human adaptation in the antarctic (in Russian), p. 208 (Medicina, Leningrad 1978).

Vasilyev, V.N.; Chugunov, V.S.: in The activity of the sympathetic adreno-medullary system at different functional states of man (in Russian), pp. 160–181 (Medicina, Moscow 1985).

Vasilyeva, I.A.: Probl. Endokrinol. 27: 81–84 (1981).

Vigas, M.: Neuroendokrinna reakcia v stresse u cloveka, p. 246 (Bratislava 1985).

Vigas, M.; Nemeth, S.; Jurcovicova, J.; Widerman, V.; Malatinsky, M.: Role of catecholamines in stress-induced growth hormone release in man; in Usdin, Kvetnansky, Kopin, Catecholamines and stress: recent advances, pp. 573–578 (New York 1980).

Viru, A.: Scientific notes of Tartu University (in Russian), vol. 368, pp. 20–29 (1975).

Viru A.: Hormonal mechanisms of adaptation and training (in Russian), p. 156 (Nauka, Leningrad 1981).

Viru, A.; Tendzegolskijh, Z.L.; Karelson, K.M.; Alek, K.V.; Smirnova, T.A.: Vzaimootnosheniia beta-éndorfina i riada gormonov v krovi vo vremia myshechnoi raboty. Vopr. Med. Khim. 33: 28–32 (1987).

Vlasova, N.V.; Gitelzon, I.I.; Okladnikov, I.N.: Lipidnyi obmen u korennykh zhitelei Krainego Severa. Vopr. Pitaniya Sept./Oct. 5: 53–56 (1975).

Volovich, V.G.: Man in the extreme conditions of the natural environment (in Russian), p. 223 (Mysl, Moscow 1983).

Vora, N.M.; Williams, G.A.; Hargis, G.K.; Bowswer, E.N.; Kawahara, W.; Jackson, B.L.; Henderson, W.J.; Kukreja, S.C.: Comparative effect of calcium and of the adrenergic system on calcitonin secretion in man. J. clin. Endocr. Metab. 46: 567–571 (1978).

Watson-Whitmyre, M.; Stetson, M.H.: Simulation of peak pineal melatonin release restores sensitivity to evening melatonin injections in pinealectomized hamsters. Endocrinology. 112: 763–765 (1983).

Wilkerson, J.E.; Raven, P.B.; Bolduan, N.W.; Horvath, S.M.: Adaptations in man's adrenal function in response to acute cold stress. J. appl. Physiol. 2: 183–189 (1974).

Wills, E.P.; Wilkins, E.A.: Release of enzymes from lysosomes by irradiation and relation of lipid peroxide to enzyme release. Biochem. J. 99: 238–251 (1966).

Wilmore, J.H.: The use of actual, predicted and constant residual volumes in the assessment of body composition by underwater weighing. Med. Sci. Sports *1:* 87–90 (1969).

Wyndham, C.H.; Plotkin, R.; Munro, A.: Physiological reactions to cold of men in the Antarctic. J. appl Physiol. *19:* 593–597 (1964).

Yakimenko, M.A.: Theoretical and practical problems of thermoregulation (in Russian), pp. 34–44 (Ashkabad 1982).

Yakimenko, M.A.; Simonova, T.G.: Loss of moisture in breathing at cold adaptation of man; Kaznacheev, Derjapa, Turchinsky, Physiology and pathology of human adaptation in the far north (in Russian), pp. 86–87 (Medical Academy of SSSR, Novosibirsk 1977).

Yoshimura, H.: Acclimation to heat and cold; in: Essential problems in climatic physiology, pp. 61–106 (Nankodo, Tokyo 1960).

Yoshimura, H.; Iida, I.: Studies on the reactivity of skin vessels to extreme cold. Jap J. Physiol. *1:* 147–159 (1950).

Zaidase, I.; Bessman, S.P.: Diabetes mellitus: etiopathogenesis and metabolic aspects, pp. 77–92 (Karger, Basel 1984).

Zoloev, G.K.: The role of parathormone, calcitonin and glucocorticoids in disturbances of calcium balance and its correction at postagressive states; Med. Sci. diss Tomsk University (1983).

Zoloev, G.K.; Slepushkin, V.D.; Kanskaia, N.V.; Avdienko, V.N.; Vasiltsev, I.S.: Funktsiia parashchitovidnykh zhelez i obmen kaltsiia pri infarkte miokarda. Ter. Arkh.*55:* 33–36 (1983).

Zoloev, G.K.; Mordovin, V.F.; Slepushkin, V.D.: Znachenie beta-adrenergicheskoi sistemy v reguliatsii fundtsii kal tsiireguliruiushchikh zhelez. Probl. Endokrinol. *33:* 60–64 (1987).

Subject Index

Acclimation 2, 3
Acclimatization 3, 9, 14, 16–19, 71
Acid
  DNAase 159, 160
  lysosomal hydrolases 158–161
  phosphatase level 159, 160
ACTH, see Adrenocorticotropin
Activation energy 165, 166
Adaptation 3, 37–41
  to cold 2, 60, 61, 63, 64, 114, 170
Adrenal cortex 115
Adrenaline 108–114
Adrenocorticotropin 11, 115–118,
  131–134, 138–140
Aerobic power 48, 49
Alarm reaction 17
Albumins 142
Aldosterone 50, 116, 119
Angiotensin-II 119, 121–123
Angiotensin-converting enzyme 121, 123
Anthropometric measurements 44, 46–48
Anthropometry 5
Apo-A 148, 151
Apo-B 151
Apoproteins 148, 149, 151
Arterial pressure 57, 58
Ascorbic acid deficiency 3
ATPase 169
Atrial systole 34

Ballistocardiogram 35–37, 41
Bazette index 33, 35, 41
Biologically active compounds and their
  status 107–138
Blood
  analysis 31
  glucose 136–138
  lipid profile 143–152
  pressure 49, 50, 56

Body
  cooling 9, 65
  fat 12
  mass 46, 47, 54
  temperature 59–63
Bronchospasm 9

C-peptide 130, 131, 141
Calcitonin 129, 130
cAMP 116, 134, 135
Cardiac cycle length 37–40, 42, 69
Cardiac rhythm indices 38, 42
Cardiolipin 151, 154
Cardiorespiratory function 5, 67
Cardiovascular function 33–43, 45, 46,
  48–55
Catecholamines 137, 138, 159
  release 107, 108, 110, 112, 113
Cathepsin levels 8
Cathepsin-D 159, 160
Cattel's 16PF test 74–77, 82, 85, 90, 94,
  101
Central nervous system
  dysfunction 18
  simulations 67–70
cGMP 134, 135
Cholesterol 111
Chylomicrons 146
Circumpolar regions 2
Circumpolar stress 13–18
Climatic chamber simulations 59–70, 149,
  150, 157
Cold
  adaptation, see Adaptation to cold
  environments 1
  exposure 5
  hormones 11
  injury 29, 30
  receptors 114

Subject Index

Cold (cont.)
  sensitivity 18
  stress 9
Cold-induced vasodilatation 60–67
Color discrimination 70
Combined stress 1
Cortisol 11, 52, 115–118, 138–140
Cramps 30
Cross-country skiing 2, 7
Cycling 45, 49

Dehydration 47
Diarrhea 25, 30
Diastolic blood pressure 58
Diet 3
Diphosphoglycerate 156
Diurnal rhythms 15
DOPA 111, 112
Dopamine 111

Electrical axis 33, 34
Electrical conductivity 161–166
Electrocardiogram 33, 34, 40, 68
Emotional stress 72, 73
Emotions 26
β-Endorphius 11, 117, 131–134
Energy expenditures 7, 8, 67, 170, 171
Enteritis 25
Ergometry 45, 55–57
Erythrocyte
  membranes 161–169
  shadows 161–169
Erythrocytes, glycolysis 154–158
Estradiol 123, 124
Estriol 123, 124
Exercise 1
  hormones 11
Expert appraisal 88, 89
Eye irritation 30

Fast-wave sleep 18
Fat mobilization 12
Finger temperature 66
Fluid balance 171
Free fatty acids 143–145
Frostbite 5, 8, 9, 26, 27, 30
Fructose-1,6-diphosphate 156, 157
Functional level of system 69

Galactosidase 159, 160
Gastrointestinal discomfort 30

Geomagnetic radiation 3, 4, 10
Globulins 142
Glucagon 130, 131, 159
Glucocorticoids 136, 141, 159
Gluconeogenesis 142
Glucose 111, 156, 157
Glucose-regulating hormones 139–142
  stress 135–138
Glucose-6-phosphate 156, 157
Glycolysis 12
Grip strength 46
Group dynamics 99, 100
Growth hormone 116, 128, 129

Habituation 2
Hand temperature 64
HDL 146–148
Heart rate 41, 42, 46, 50–53, 67
Heat generation 16
Hemispheres, brain 18
Hemodynamic indices 36, 42
Hemoproteins 153
Hexokinase 155
5-HIAA 111, 112
Histamine 110
Histidine 110
Hormonal balance 1, 2
Hormonal changes 11
Hormones and their status 106–138
5-HOT 111, 112
HVA 111
Hydrostatic weighing 44
Hypertension 35
Hypochondriasis 99
Hypothalamus/hypophysis/gonadal
  axis 123, 124
Hypothalamus/hypophysis/thyroid
  axis 124–127
Hypoxia 13

Ice 24
Infrared spectroscopy, membranes 163, 166–169
Insulative reactions 16, 17
Insulin 130, 131, 136, 138, 141
Insulin/C-peptide/glucagon system 130, 131

Kerdot's autonomic index 35, 36

Subject Index

Latitudes, high 16–18
  stress 71, 72
LDL 146–148
Lecithin-cholesterol-acyltransferase 146, 147
Level of functional possibilities 69
Lipid
  metabolism, stress 135–138
  peroxidation 10, 152–154
Lipids 136–138, 143–148
Lipolysis 135
Lipoprotein lipase 146, 147
Lipoprotein-2 cholesterol 12
Lüscher's color test 76
Luteinizing hormone 123, 124
Lysoforms 154
Lysosomes 158–161

McRause index 33, 34, 40, 41
Malone dialdehyde 153
Metabolic acclimatization 16
Metanephrine 110
Minnesota Multiphasic Personality Inventory 74–77, 82, 85, 90, 94, 101
Motivations 96–98
Muscle strength 44, 46–48
Muscular pains 30

Neuromediators 106
Neuropeptides 131–134
Neurotensin 132–134
Noradrenaline 108–114
Normetanephrine 111
Nutrition 5, 11, 12

Osmolality 119–121
Oxygen
  cost 52–54
  intake 7, 8, 45, 49, 52, 53
Ozone, hormones 11

P wave 33, 40
Parasympathetic tonus 15, 33, 34
Parathormone/calcitonin system 129, 130
Parathyroid hormone 129
Peer appraisal 88, 89, 103, 105
Peroxidation 152–154
Personality traits 91–93
Phagocytes 159
Phase transition 161, 163, 165
Phosphatidylethanolamine 151

Phosphofructokinase 155
Phosphoglycerate kinase 155
Phospholipids 144, 148–152, 166
Physical health 29
Physical working capacity 45, 57
Pituitary/adrenocortical axis 115–118
Polar day 26, 27
Polar dyspnea 17, 18
Polar stress syndrome 15, 16
Potassium 117, 119–121
PQ interval 34, 40
Progesterone 117, 118
Prolactin 123, 124
Prostacycline 135
Prostaglandins 134, 135
Psychic adaptation, cold 71–105
Psychoemotional stress syndrome 71
Psychological changes 18
Psychomotor changes 18
Psychophysiological function 5
Psychosociological function 5
Pulmonary function 44
Pyruvate kinase 155, 156

QRS interval 40, 41
QT interval 35, 41
Questionnaires and psychic adjustments 74, 75

Radiation 30–32
Radioimmunoassay 107
Raven's progressive matrices 75, 76, 79
Reaction stability 69
Rectal temperature 59–63
Renin 119, 120
Renin/angiotensin/aldosterone axis 118–123
Respiratory rate 67
Respiratory reactions 17, 18

Searching activity 98
Self appraisal 88, 89, 104
  mood 68, 75, 85, 88, 90, 92, 100
Serotonin 109–113, 133
Serotoninergic system 108–114
Serum
  lipids 143, 144
  lipoproteins 144, 145
  proteins 142–144
  triglycerides 144, 145

Subject Index

Shivering 63
Simulation, ski trek 59–70, 148, 149
Sinus arrhythmia 34, 35, 40
Ski treadmill 45, 52, 53
Skin
  damage 30, 31
  temperature 59–63
Skinfold test 46
Slow waves 36, 37, 40
Sodium 117, 119–121
Somatomedin-C 117, 127–129
Somatostatin 117, 127–129
Somatotropin 127–129
Strain diabetes 141, 158
Stress 13–18, 36, 37, 41, 42, 69, 71–73, 136–138, 142, 170, 171
  index 39, 42
  physiological 1–3, 10, 11, 14, 15
  psychological 14
Subjective appraisals 75–91
Substance P 132–134
Sunburn 2
Sympathetic activity 36
Sympathoadrenal system 108–115
Systolic blood pressure 49, 50, 56, 58
Systolic index 35

T_3 125–127
Temperature measurements 59–63

Testosterone 123, 124
Thermal regulation 109
Thromboxane 135
Thyroid hormones 141
Thyroxine 124–127, 141
Tocopherol 153
α-Tocopherol deficiency 3
Toe temperature 66
Total lipids 111
Triglycerides 111
Tryptophan 111, 112
TSH 125, 126

Ultraviolet radiation 3, 4, 10, 31
Upper body strength 47
Urinary proteins 142, 143

Vanillylmandelic acid 110
Vasoconstriction response 64, 65
Vasopressin 119–122
Vascular changes 17
Viscosity measurement 161, 163, 164
Vitamin E 153
VLDL 146–148

Windchill 8, 9